W9-DHS-640

THE SCARS OF PROJECT 459

The Environmental Story of the Lake of the Ozarks

TRACI ANGEL

The University of Arkansas Press
Fayetteville
2014

For those who care about the Ozarks region
And to my future: Jason, Jack and Earth Day baby, Zara

CONTENTS

ACKNOWLEDGMENTS

Thank you to H. Dwight Weaver for his ear, advice, photos, and overall knowledge of Lake of the Ozarks history and matters; Randy Miles, Tony Thorpe, Bob Broz, Dan Obrecht, and Jack Jones at the University of Missouri; Donna Swall for her access and availability to the Lake of the Ozarks Watershed Alliance; Caroline Toole, Cindy Hall; Bill McCafree; Ken Midkiff for his candidness; and Steve Mahfood and Earl Pabst, who restored my faith in people who will still talk without fear.

Those who added a tip or suggestion along the way: Kathleen Logan Smith, Ted Heisel, Dave Murphy, Melinda Hope, and Peter Goode.

For help with fact-checking the science: Rebecca O'Hearn, John Schumacher (and for allowing me to accompany him on an educational field trip), Ken McCarty, Tracy Rank, and James Vandike.

Barbara Fredholm, who took me on a kayak adventure and assisted with local information; Celeste Miller, Steve Spalding, Ellen Harrel, and Nanci Gonder; Vickie Bailey of Schultz Survey and Engineering, and Stan Schultz; Mary Reagan in Greg Gagnon's office, and Greg Gagnon; Jim Divincen, Arlene Kreutzer, Joe Pryor, Joe Roeger, and Danny Roeger; Pat Shellabarger in Camden County Planning and Zoning; Brent Salsman in the Camden County assessor's office; Mickey and Michael McDuffey; Chad Livengood; the Missouri State Archives; the Missouri Historical Society at the University of Missouri–Columbia; and Stuart Westmoreland of the state Department of Natural Resources.

Warren Witt, who spent a morning speaking to me about utility company Ameren's modern commitment to the lake; Alan Sullivan; Marvin Boyer of the US Army Corps of Engineers; Kris Lancaster and Emily Barker of the Environmental Protection Agency; Sharon Harl, Dan Field, Mark A. McBride, and Mike Gillespie; mentors Michael Grinfeld, Jacqui Banaszynski, and Bill Allen for continual support and teaching; Grazia Svokos and Thomas Crone for their sound advice before I signed the contract; Elaine Adams at the *Kansas City Star*, for keeping me with regular freelance income and offering encouragement along the way; copy editor Pamela Hill, Larry Malley, and the University of Arkansas Press staff for taking a chance to see this book come to life.

To friends who asked about the book and gave feedback, including the women of Kansas City's Best Book Club Without a Good Name and of Again With the Book Club, and Sarah Reese, a great friend, and lucky for me, an excellent copy editor.

To my family, for their support: my parents, Glenda and Jim Angel, and my in-laws, Roger and Carol Morrison, for watching the kids so I could report and write; Jack and Zara for enduring numerous road trips in their car seats and keeping me going with their smiles and laughter; and, my live-in scientist and partner, Jason, for believing in me, helping with last-minute edits, looking over terminology, and for his encouragement while spending all free moments and family vacations on this consuming project.

PROLOGUE

I stand near the edge of a wooden dock with bruised knees and tomboy hair spilling into a ponytail. It's the summer of 1985, and I've walked the quarter mile from our Lake of the Ozarks vacation condo to look out into the murky water and wonder what is below. This looks different than the stony creek in my backyard or the ponds I race to for refreshment.

The Lake of the Ozarks is fairly young by lake standards. It formed in 1931 when the Bagnell Dam clasped the Osage River to fill the reservoir—an undertaking dubbed Project 459 by the federal government. Its design inks Missouri's central map like a dragon tattoo, with a shoreline that sprawls into four counties with tiny villages snuggling up to its banks.

Its birth, and quite likely any signs of its death, hinges on a unique characteristic—it is privately owned and run by a utility company, Ameren, formed by the merger of Central Illinois Public Service Company and Missouri's Union Electric Company.

I have followed the Lake of the Ozarks' environmental situation as a journalist since 2006. I first read about a watershed group organizing while at the University of Missouri during a midcareer journalism fellowship. The idea that a group had formed to protect the Lake of the Ozarks made me think of that childhood vacation in which I peered over the dock into a dark abyss. I was horrified as an eleven-year-old who wanted to go swimming, but whose mother would not allow it. Over the years, I've returned to the area. The hiking and surrounding areas are gorgeous, but the lake water brought no appeal, even to a girl who has swam her share in country streams.

My first impression of the lake's problems was that it illustrated development gone awry. A cruise along the main highways reveals clusters of condominiums and houses jammed together, squeezing one another out for a space at the water's edge.

Through my research I learned that small efforts were made in previous decades to put a development plan in place to protect the watershed, which spans 8.96 million acres from eastern Kansas into central Missouri. Many lake communities used planning and zoning boards or area-wide development plans to protect the water quality and ensure the health of the area's prime resources.

In the Lake of the Ozarks' case, the stronghold of business owners and tourist-based locals deny and quickly silence any thought or expression of trouble because it's bad for tourism. This is the big vacation destination for this part of the Midwest, after all.

I wrote a five thousand–word magazine piece for *Ozarks* magazine in 2008, but the publication folded before it was published. The *Columbia Tribune* published a condensed news article for me in 2009. I have collected much material and many contacts over the years, and the story continued to haunt me.

During the reporting of this book the lake business community would ask, "Is this going to hurt us?"

I received plenty of discouragement, even from opposite ends of the political spectrum—from lake people and others with a vested business interest to the environmentalists who say the lake's environmental status is a lost cause. No one at the lake wants to admit any problems, although they are all working to keep the lake clean and make improvements—improvements on something that they say is completely fine.

No one wanted to say anything negative about the situation. They all had neighbors or coworkers whose businesses and livelihoods thrive on the tourism industry. Even the watershed group kept me at a distance, sending literature and studies my way but never acting like an objective watchdog, as it seems was its original intention. Local people had opinions about what was happening, but couldn't point to anything that might support their judgments. Some refused to go on record because they feared community ostracism and small-town backlash.

State agencies, all under Missouri's so-called environmental governor, Democrat Jay Nixon, closed up and muzzled scientists and employees when I reached out for interviews. My guess is they had concerns about how they might be portrayed after they faced criticism for hesitating to release reports of high *E. coli* counts in the water before a busy holiday weekend in 2009.

Finding that storytelling voice from a whistleblower, which often helps these kinds of stories along, proved to be more difficult than I had imagined.

Instead, I dug up the documents, newspaper articles, and historical accounts about what has happened to the lake over the decades. Development seemed to bubble up without warning.

While the lake might not be horrible or unsafe for swimmers or fish, it has problems. Discharges of sewage from failed septic systems and erosion runoff from parking lots and fertilized lawns are common.

The lake lacks the regulations of other government-formed lakes like those of the US Army Corps of Engineers, which restricts construction and zoning all around the shoreline. Yet dozens of townships, four county governments, Missouri state agencies, and the federal commission responsible for power plants all claim a stake and say when it comes to law and control of the Lake of the Ozarks.

The factors of private ownership and confusion of ruling governing bodies all play a role in the lake's current challenges of threatened water quality from failing sewer systems and pollution from the area's construction.

This book will document how years of building and perhaps relaxed regulation, or no regulation, has created a reason for a call for action. The evidence in the following pages is documented through scientific studies, public records, and thorough interviews with stakeholders.

Cited works and bibliographical materials for each chapter are noted at the book's end. I did this to allow readers a smooth transition from page to page. I used direct quotations and referenced them when necessary. All quotations from people are from first-hand interviews and correspondences unless otherwise noted. I received neither grant nor funding from anyone during the writing and reporting of this book. My only motives were to present the truth, as I could find it, to readers. I take full responsibility for any unintentional errors or omissions.

This is the story of what happens when a body of water is *loved to death* but has few limitations or restrictions as it grows. Those living beside the lapping, boat-fueled waves deny and deflect blame like protective mothers. Others look the other way.

And then there are others who wish to save it, but who are bound by obstacles created years before their involvement. They cautiously tread among the business owners and residents who make the lake's tourist industry their livelihoods.

Meanwhile, the region's largest tourist destination awaits its environmental fate.

THE SCARS OF PROJECT 459

CHAPTER 1

Nature's Attraction

Conwee, the son of an Osage Indian chief, lived where Ha Ha Tonka State Park at Lake of the Ozarks sits today.

As legend goes, he fell for the daughter of another chief who lived north of the Osage River and wanted her for his wife. But the girl, Wasena, didn't return his affections. Conwee set out for her anyway and left his camp one night to cross the Osage River. He snatched Wasena and started back to his camp. They stopped at a cave to wait until morning. Wasena broke free from her captor and ran toward a high cliff above the Niangua River. Conwee ran after her and as he approached she jumped over the edge. Wasena chose death over living her life with Conwee, whom she did not love.

That cliff at the confluence of the Osage River and Niangua River is now called Lover's Leap.[1,2]

And that beautiful bluff, with the scenic outlook over what is today the Lake of the Ozarks, could be demolished to build condos.

The Lake of the Ozarks, and its more than 1,150 miles of shoreline, spring from a map like a vacation oasis in the landlocked Midwest. Many Missourians and Arkansas natives hold the lake amid their fond memories of summer getaways, fishing trips, and family reunions.

Created from the Osage River's damming, the lake became a major tourist destination in the mid-twentieth century, drawing visitors to its quirky amusement stands and inviting waters. The lake's central area is about forty-five miles southwest of Missouri's capital, Jefferson City, and a few hours' drive from its major metropolitan areas, Kansas City (160 miles) and Saint Louis (175). It also attracts many from southern Missouri and the northern Arkansas areas.

Eighty years since the damming, however, the uncontrolled development and only recent monitoring of water quality are creating headlines and buzz about what's lurking beneath the surface. Reports of sewage spills and elevated levels of nutrients—all results of the human impact and development—have muddied the waters and scarred its pastoral image.

What's at stake is what attracts people to the Ozarks region: its natural beauty bursting from its diverse landscape no matter what the season.

"There are several parts of the state—and the Lake of the Ozarks region is one of them—where savannahs, open oak woodlands, glades, and prairies still look very much as they did two hundred years and more ago," said Ken McCarty, natural resources manager for the Missouri Department of Natural Resources.[3]

Lake of the Ozarks State Park, which spans more than seventeen thousand acres, contains several areas where park managers have restored and now preserve natural remnants which show how the landscape appeared in the eighteen hundreds, well before the lake was built.

Historically, dry oak woodlands covered the land, their trees so widely spaced that the ground beneath flourished with grasses, shrubs, and wildflowers. The high, flat plains would meet open prairies, where the tall grasses and sparsely growing trees sharply contrasted the more densely timbered landscape deep in the river valleys.

The Native Americans kept the fields open through periodic fires and burning, said Lake of the Ozarks State Park naturalist Cindy Hall.

They cleared the land to ward off enemies, corral animals, or create a wider space for hunting. On horseback, the natives could see farther into the distance when the land had fewer woodlands and brush.[4]

Bison and elk roamed where the tourists now trek to their condos. Wolves stalked their prey in the woods that now lead down to the muddy waters. Many species of butterflies floated among the prairie grasses.

The fires killed young saplings that popped up amid the prairie species. Having periodic fires developed the region through bluestem and other brush.

This overlook that allows views of the Niangua and Osage Rivers is known as Lover's Leap. The iconic spot has lured many to snap of photo of the view. The property was zoned for condominium development in 2005. (*Photo courtesy Mark A. McBride.*)

Two women sit on the rock at the Lover's Leap viewing spot, off Highway 5, in the 1940s or 1950s. (*Photo courtesy Missouri State Archives.*)

When the deliberate burning ceased, the tree canopies closed out the sky. It created a chain reaction for the flora and fauna that relied on the wide-open spaces for their survival.

The damming in 1931 created a large body of water that replaced the constant moving of the Osage River with much more oxygenated water to sustain different life.

"It switched to a lake that is deeper and darker with slow-moving water," Hall says.

Most of the changes in the last one hundred fifty to two hundred years are from humans. However, everything transformed so fast that other species could not adapt or change, Hall says.

It's the birds that are the most significant sign of what is lost and what can be lost. Grassland birds and the migratory birds are without homes. Some species are still around, but their numbers and sightings are so rare they are on protective lists. Red-headed woodpeckers still find homes in the Ozarks, but are a disappearing sight and bird experts are keeping tabs on populations there.

"If the number crashes, it could have detrimental effects on [woodpecker populations] in North America," Hall says. "So many require a large, uninterrupted space, and you just don't have that anymore."

The gray bat and Indiana bat were also listed on the federal and state endangered species lists.

Coakley Hollow Trail in Lake of the Ozarks State Park passes by a boggy wetland and then dolomite glades, often referred to as Missouri's deserts.[5] Riddell's goldenrod, water-tolerant rushes, sedges, and crayfish live on the wetlands, just downhill from prickly pear cactus, drought-adapted yellow coneflowers, Missouri primrose, and scorpions on those glades, McCarty said.

In other areas of the park, flowering dogwoods appear in springtime, and asters, white oaks, and bluestem poke through during the fall.

Yellow coneflower, standing slender and golden, endemic in the world only to the Ozarks and throughout the Ha Ha Tonka glades, are finding their growing areas are disappearing outside of managed natural areas and parks.

Exotic plants, such as the *Sericea lespedeza*, and the bush honeysuckle, have crept in and colonized, squeezing out other plants. *Sericea lespedeza* was once used in erosion control along roads, but it quickly spread to hay fields, glades, and prairies and then took over.

"It was once planted on roads as cover and in fields for wildlife browse and livestock forage, but now it is considered one of Missouri's most invasive weeds," McCarty said.

Ha Ha Tonka State Park is the area's natural jewel. Within the two thousand seven hundred acres are some of the state's most scenic nooks and sites. Caves, fifteen miles of trails, and the ruins of a twentieth-century castle lure hikers to traverse the area.[6] Besides its beauty, lore, and legend, people from the Native Americans to the wealthy Snyders, a prominent family from Kansas City, have called it home.

One trail, Colosseum, weaves below a natural bridge and through a sinkhole from which the trail is named. A stone arch that forms the bridge is what remains after the caves surrounding it collapsed.

Another hike, Spring Trail, winds by the Lake of the Ozarks shore-line and Hahatonka Spring. A millpond along the way is home to frogs and green herons. Another overlook commonly offers views of muskrats and otters.

Ha Ha Tonka is an example of public land that has preserved Missouri's distinctive landscape. The park contains one of the area's best examples of its historic open woodland and glade landscape.

People since prehistoric times, including the current park managers, burned these woods and kept this landscape open enough to preserve these native environments, said McCarty.[7]

"We have been implementing prescribed fires to restore the historic character and inherent richness of plant and animal life," McCarty said.

The fires have helped the area retain much of the biological diversity and its robust mix of plants and species. A person who spends enough time in the park can find more than eight hundred different species of plants, he said.

Fire that the first settlers used to help keep the woods open enough to accommodate their free-ranging livestock also helped preserve many of the native plants and wildlife, which is an important goal in state parks today, McCarty said.

The importance of those efforts in state parks, and on public and private lands, is evident with what happens to plants and wildlife when these natural landscapes are lost or fragmented. Many need a certain

Ha Ha Tonka State Park, considered one of Missouri's most beautiful, features the ruins of a castle that was built by businessman Robert Snyder. The trails weaving through the park note the area's distinct geological features, such as sinkholes, caverns and springs. (*Photo courtesy of Missouri Department of Natural Resources.*)

amount of space, and if it becomes broken into too many patches, they can no longer live in the area, he said.

As the lake's natural communities fall prey to development, man-made growth preempts space that plants and animals require.

"The ability to preserve rich and ecologically healthy natural environments becomes more challenging, and it also becomes more significant, with expanding urbanization," McCarty said.

CHAPTER 2

History of the Lake

Before there was a lake, there were the Osage Indians. They traveled along the river which now bears the tribe's name, and climbed its surrounding bluffs stemming from the area's roller coaster dips and valleys carved from erosion and land lifting. They wandered and inhabited the region's cave systems. Many of the area's names are taken from the Osage, such as Ha Ha Tonka, thought to mean "laughing waters" in the Osage language.

Other notorious travelers were Daniel Boone and his son, who hunted the area woods. By the early eighteen hundreds, others came to find their livelihoods and homes near the rivers.

The settlers pushed the Osage into Oklahoma before they put down their roots. They cultivated the land surrounding the river, shipping wheat to nearby mills. They navigated their steamboats past river bluffs and landmarks with iconic names—Bloody Island, Bat Cave, Lover's Leap.[1]

Their river was used for trade and for powering sawmills or woolen mills that had become part of the region's infrastructure.

By the early nineteen hundreds, a population of more than fifty thousand people[2] lived in what is now the four-county (Benton, Camden, Miller, and Morgan) area surrounding the river, which finds its beginning in the Flint Hills as the Marias des Cygnes in eastern Kansas and becomes the Little Osage near Rich Hill, Missouri. The waterway continues east and joins the Missouri River.

These people who called this region home would soon find themselves uprooted and without a land when a power company from Saint Louis decided to cash in on a hydroelectricity project.

BAGNELL DAM TIMELINE

August 6, 1929	Construction begins
Nov. 21, 1929	First steel setting on Osage bridge
April 9, 1930	First concrete poured
July 22, 1930	Diversion channel opened
Feb. 2, 1931	Lake begins to fill
May 20, 1931	Lake reaches spillway crest elevation
May 30, 1931	Lake traffic begins
Oct. 16, 1931	First commercial operation of Osage plant
Dec. 24, 1931	Lake area electric service begins

Source: Miller County Historical Society

The damming of the Osage River for hydroelectric power, which would ultimately create the Lake of the Ozarks, began amid financial corruption.

Though the idea for Bagnell Dam came from Kansas City attorney Ralph W. Street, it was businessman Walter Cravens of Kansas City who formed the Missouri Hydro-Electric Power Company and offered the financing for damming the Osage in 1924. Cravens was the first president of Kansas City Joint Stock Land Bank.

The company began building roads and created facilities so the dam could be built near the town of Bagnell. Then financial difficulties halted construction in 1926.[3] Authorities investigated Cravens's management of the bank's assets, including questionable relationships with Farmers Fund Inc., the company charged with acquiring lands in the Osage River valley. Bankers started pulling out of the project. A year later, Cravens was indicted for phony loan activities.[4, 5]

However, others continued to pitch the idea and negotiated a sale to another company. In 1929, Union Electric Light and Power Company in Saint Louis announced it would build the dam for $30 million.[6] The company previously had purchased a steam-power plant in the village of Rivermines.[7] They started building the Bagnell Dam so they could meet power needs. The development was another addition to the company's other hydroelectric ventures along the Mississippi River.

Shortly after, the New York Stock Exchange crash spiraled into the Great Depression.

People poured into the area for the work. One estimate is that twenty thousand people slept and ate in makeshift camps during the two years it took to build the dam. Workers labored for as low as thirty-five cents an hour. Cottages, bunkhouses, and even a bakeshop created a temporary town for the workforce.

Dam construction jobs were welcome blessings for many looking for work during the Depression, and they signed up, no matter how laborious the work.

Alan Sullivan, consulting engineer at Ameren, likes to tell the story of his family's connection to the lake.

"In our little farm town, my uncle and grandfather were at Sunday night church . . . " he begins. He tells the tale matter-of-factly, as he has many times before. He's proud to come from a line of descendants from the area. The country was sliding into the Depression, and the two men had heard that Union Electric would be hiring to build the dam.

"They rode horseback most of the night to stand in the hiring line," Sullivan says. Many people stood there, waiting anxiously, with tensions mounting. A fight broke out near Sullivan's grandfather and, as the story goes, his grandfather's respectful avoidance of the confrontation helped him land a job. His grandfather said he would only work if they gave his son a job, too.

They were hired for thirty-five cents an hour and worked for two years.[8]

Though the dam's building created jobs, it meant changing many things about the landscape. A lot of cemeteries were moved prior to the flooding. People left their homes for lives elsewhere, and entire villages drowned beneath the water as the lake filled.

Joshua William Vincent, the publisher of the *Reveille* in Linn Creek, took a stance against Union Electric as the dam project began.

His column in March of 1928 expressed his views.[9]

> Some believe that money cannot compensate for destruction of some of the world's choicest beauty spots to make way for a rotting fester in the heart of our state which even time cannot cure.

He called for people to take action.

> Whatever the differences, then, can any object or interest be served by inaction? We are here, we belong here, our interests are here; why not make ourselves known?

Workers lived for weeks in camps near the dam's construction. *(Photo courtesy of Missouri State Archives.)*

Building the Bagnell Dam put thousands of people to work during the hard financial times of the Great Depression. *(Photo courtesy of Missouri State Archives.)*

Construction on the dam began in 1929, but the lake didn't begin filling until two years later. The area became a focal point for the tourism industry that is the main economic driving force of the Lake of the Ozarks today. *(Photo courtesy of Missouri State Archives.)*

The people of the Ozarks region now love the lake, say authors Leyland and Crystal Payton in their 2012 publication, *Damming the Osage: The Conflicted Story of Lake of the Ozarks and Truman Reservoir*. But a chat with old-timers or a read of the *Reveille*'s back issues reveal apprehension and perhaps even suffering from the development, the Paytons write in their book.[10]

The damming process moved an estimated one million cubic yards of earth and cleared thirty thousand acres of trees, bushes, and green space. About sixty thousand railcar loads of materials were hauled to the site. One estimate is that the concrete used to build the dam is the equivalent of constructing an eighteen-foot roadway from Saint Louis to Topeka, Kansas.

In the end, a concrete gravity dam measuring 2,543 feet long towered above the river. Within its walls was a 511-foot-long power station. When the turbines are off, seventy-ton steel headgates seal the openings.

When the turbines are on, they revolve with a force equal to 33,500 horsepower each and produce a maximum of 21,500 kilowatts each. Today, in a typical year, the company estimates the Osage Energy Center, as it is called, produces more than five hundred million kilowatt-hours of electricity to support nearly forty-two thousand average households.[11] The building of the dam brought electricity to many of the rural areas that did not have it before.

Another result of the damming was the impact on the floodplain. The created project had a projecting point on the south side of the floodplain. A sloping ridge was on the north side, and the positioning helped restrict the floodplain valley more narrowly than before. It was scaled back to one-half mile in width. Under the modern licensing agreement with the Federal Energy Regulatory Commission, Ameren is required to adjust the flow from the "Osage Project to alleviate downstream flooding on the Osage and Missouri Rivers."

In 1931, the water started gushing and the lake began to fill. From the backwaters of other rivers, including the Pomme de Terre, Grand, and Niangua, it came. What resulted was a reservoir covering more than fifty-five thousand acres. The lake is just more than a hundred feet deep at its deepest point, and its full elevation is 660 feet above sea level.

By late May of 1931, the bridge above the dam was open for cars carrying visitors who flocked to the area.

Just as devastating was what happened to the people who lived in the tiny villages that dotted the landscape. The lake's creation swallowed towns such as Cape Galena, Proctor, Purvis, Zora, and Old Linn Creek.[12, 13]

Small though the towns were, water from the Osage River's cinching drowned out most or all of them, sending the inhabitants fleeing to find work and livelihoods elsewhere.

Many of those uprooted were poor farmers without anywhere to go. They took the small settlement Union Electric offered them—some estimates put the settlement offers at thirty-five dollars an acre—and attempted to make a life in surrounding counties.

A more notorious and wealthy family was also affected when the waters came rushing over the area. Their standoff with the utility company provides a prominent example of how the lake affected those who lived in the region.

Robert McClure Snyder, a prominent Kansas City businessman, had purchased much land around Ha Ha Tonka after seeing its beauty. The Snyder family built a half-million-dollar mansion on an estate in the manner of a European castle atop the 250-foot bluff overlooking Hahatonka Spring's branch, one of the most beautiful settings in the region.

The utility company offered the family just over $140,000 for surrounding easements, but the Snyders wanted more and sued. The family accused the company of infringing upon the scenic beauty of their home by building the lake. They even brought in a famous witness—Gutzon Borglum—a sculptor of Mount Rushmore, who spoke about Ha Ha Tonka's beauty and estimated the value at $1.5 million.

A jury found in favor of the family and awarded them $350,000, but years of costly appeals reduced the amount to $200,000.[14]

The Federal Power Commission report released a document more than a decade after the lake formed noting Union Electric bribes and kickbacks to politicians and other costs during the building. The Federal Power Commission, formed in 1930, was replaced in 1977 by the Federal Energy Regulatory Commission.

In August of 1941, Union Electric was selling two private lodges used to entertain lawmakers and politicians, according to an article in the *Kansas City Star*.[15] Real estate agent Harry Williamson said they were for sale because "the new management has changed the company's policy and has no use for them." The company closed the lodges when the US Securities and Exchange Commission investigated the company and its officials, the *Star* article said.

In 1943, the Eighth US Circuit Court of Appeals denied a rehearing of the indictment against Louis Egan, Union Electric's president, on a charge of conspiracy to violate the Public Utility Holding Act of 1935.[16]

The law quoted in the verdict stated it was unlawful for holding companies to make "any contribution whatsoever in connection with the candidacy, nomination, election, or appointment of any person for or to any office or position in the Government of the United States, a State, or any political subdivision."

The indictment alleged a secret cash fund was created that could be used for making political contributions.

After the dam was built, the Osage River area, which had once been a place for farmers, millers, and miners, was ready to entice visitors to a tourist destination with a sinuous body of water that welcomed vacationers from the landlocked Midwest.

A week after the dam was opened in May 1931, visitors gravitated to Eldon and a vacation spot was created. Shortly afterward, a *Time* magazine article from December 14, 1931, made an eerie prediction about its increasing popularity.

> A dragon will crawl across future maps of Missouri. Its head will be at Bagnell, in the central part of the State. Its tail and claws will twist 129 mi. westward. It will be labeled Lake of the Ozarks, largest artificial lake in the world, created by a dam across the Osage River built by Stone & Webster for Union Electric Light & Power Co. (North American subsidiary). St. Louisans and Kansas Citizens will have summer shacks and duck shooting lodges along its 1,300 mi. of shoreline. St. Louisans, who will consume a large part of the dam's annual 425 million kilowatt hours of electricity, also hope that Lake of the Ozarks will temper their city's blistering summers.[17]

Lasting effects of the Great Depression and then later World War II halted the area's growth to some extent, but signs of its future expansion began to emerge. While many families moved away, others stuck around, such as Alan Sullivan's and longtime lake fixture, Buford Foster, whose family was displaced from Linn Creek but ended up acquiring much land in the lake area.[18]

By the early 1940s, roadside eating places and novelty stands greeted motorists. Then came the resorts and lodges, followed by homemade food and candy stores and carnival attractions. Some of the smaller nearby towns gained business from the new tourism industry.

In his book, *Lake of the Ozarks: The Early Years*, H. Dwight Weaver describes how the town of Eldon profited.

"Eldon was transformed from a country town to a bustling little city," Weaver writes.

Camdenton, which became the county seat in 1930, transformed from part of the Chapman farm to a populated area, Weaver writes.[19] The town counted seventy dwellings, four general stores, an airport, bank, two cafes, and two hotels, among other amenities.

While the tourists flocked to the area's novelty shops and relaxed atmosphere, other markets were still trying to find a rebirth after the lake changed the locals' occupational focus.

The *College of Agriculture Bulletin*, a University of Missouri publication, in 1942 described how "had it not been for the great bulk of really good lands of the lake area were flooded upon the formation of the lake the benefit to the farming of the area would have been greater. This flooding of good farming land may be a matter to be regretted."

However, the bulletin article continues that the intended use for the area was for a purpose considered more valuable than farming. Perhaps it was the best decision for handling suboptimal farming land, however, since farm prices were low at the time, the article states.[20]

The article concludes with a caution what could happen in the years ahead.

"Development so far has been hit and miss with no shadow of a central plan for guidance," the report says. It concludes that the state General Assembly should enforce rural zoning that would enable needed public control over the area's development.

Though the bulletin was published just a decade after the lake opened to tourists, the report rang with an eerie foreshadowing that brings us to its present environmental situation. In this modern time fingers point to government agencies, developers, and years of head-turning as blame is placed for any ecological damage caused from a failure to plan ahead for development.

Union Electric, now called Ameren, never had and still today doesn't have a plan for the Lake of the Ozarks when it comes to development and growth.

"We are not authorized, it's not expected of us, and the community doesn't want it," said Ameren hydro-operations manager Warren Witt, in the summer of 2012.[21]

Ameren can regulate only the shoreline, and the company has limited ability to do even that, he said.

In the big picture, the company can regulate the docks on its property, and deals with anything crossing onto Ameren property, such as wastewater-discharge permits, on a case-by-case basis.

Today the miles of shoreline beckon visitors and lake dwellers to the wooded lake community that includes multiple counties and the larger incorporated towns of Versailles, Ozark, Sunrise Beach, Osage Beach, Eldon, Gravois Mills, and Camdenton. Because it's not a flood-controlled lake, regulated by the Army Corps of Engineers, most of the waterside property is privately owned. Construction continues, and an estimated seventy thousand residences dot the coastline. There are about twenty-five thousand boat docks and in some places they seem to choke the coves. The population is expected to escalate, as baby boomers in Missouri and throughout the Midwest pick an Ozark retirement locale.

On a Friday morning in August 2006, H. Dwight Weaver sat in a handmade wooden chair at Willmore Lodge visitor center in Lake Ozark and looked out onto the water, still calm at midmorning. Workers completed the lodge in 1930 for use by Union Electric, sturdied by white pine logs from the Pacific Northwest.[22]

"It gets busier as the day goes on," Weaver said.

Willmore is a cabin in the woods for which a naturalist would pine. Visitors stop for an informal visit to the lodge museum and then walk out to lean against the wrought iron fence on the deck that gives one of the lake's best views. Yet, visitors don't have to go far to see the utility company responsible for forming the lake. The Ameren Osage Power Plant has an overlook for visitors just across the highway.

Weaver wears a nametag because he gives a lot of lectures on the history of the lake area. He wears a hearing aid and carries a cell phone on his belt. He's been on a mission for several years now to preserve and record the lake's history. He had written seven books covering topics about the Lake of the Ozarks region by 2012, and a search on the Internet for any lake history sends a viewer to one of Weaver's publications. Weaver has visited nearly every foot of shoreline and photographed businesses, condos, and the lake's outstanding natural features.

Weaver points toward regions of the lake as he explains the history. He grew up in Jefferson City and came to the lake to fish with his dad in the early 1950s. One of the perks was to get to visit Dog Patch on "the strip." It was a popular store on the main business road leading off the dam where there were reptiles behind glass and a lion in a cage. Boat rides at Loc-Wood Dock were another treat. He later made his home at the lake with his wife in 1964. They joined the Camden Historical Society and started collecting Lake of the Ozarks memorabilia as a way to preserve images of the past. He worked fifteen years for the Missouri Department of Natural Resources' Division of Geology and Land Survey and was the public information officer when he retired.

Perhaps no one knows the Lake of the Ozarks area and history better than author H. Dwight Weaver, who has prolifically preserved the region's pastime in his books. *(Photo courtesy of H. Dwight Weaver.)*

As the sun creeps higher and parasailers float above waves and speedboats slice the water, it's clear that the weekenders have arrived.

"There's a lot of mixed feelings with regard to the influx of people," Weaver says. "Everybody who moves to the lake wants to be the last person to move here."

CHAPTER 3

Sounding the Alarm

Heavy clouds hanging low will bring drenching rains from the heavens, and every hillside gully, friendly creek and singing rivulet must add their bit to the rising waters. So long unshackled and free, they will dash fretfully against the great retaining wall, pitting their impatient strength against its indomitable surface.

But in time the leashed giant will gradually cease its struggles to be free; the Lake will settle into silent beauty at evening homeward winding birds may pause to drink; boats will slide hither and yon upon its glassy surface while the unrivaled beauty of green-clad upland and whispering trees that guard its coasts will be refleted [sic] in the quiet depths.

And while this sleeping giant will be the stronghold of dynamic forces that must go forth to electrify the busy wheels of great industries, many tired and brain-weary city folk will find soothing rest and balm for tortured nerves in the quietude and picturesque beauty that surrounds the new Lake of the Ozarks.[1]

—MARCH 15, 1932,
LAKE OF THE OZARKS NEWS

Early-morning fog hugs the hills of the roads leading to the Lake of the Ozarks and Osage River valley.

Only a few boats will cut wakes behind them today. It's January 2012, the off-season, and only the residents with regular jobs travel the highways that compose the lake's main drag.

This morning the movers and shakers and the just curious are headed to a Camdenton Chamber of Commerce breakfast with a presentation from Ameren about the shoreline-management details that have snagged newspaper headlines and created many headaches for Ameren's spokesperson Jeff Green. Green, in turn, coordinated town hall and public meetings to keep peace and answer growing concerns.[2]

The problems started after Ameren submitted a shoreline-management plan to the Federal Energy Regulatory Commission (FERC) to address the lake's environmental, recreational, historical, and other concerns. It's part of the process to relicense the company.

Part of the plan was the redrawing of property lines that could mean the removal of nearly 1,500 homes if the boundaries were redrawn because Ameren actually owns the land on which they were built.

That means a lot of fired-up homeowners who purchased and built homes over the decades want to know how it is they've lost the right to the property they thought they owned.[3] Missouri's congressional representatives have stepped into the fight. The state Department of Natural Resources also is involved.

And what the battle highlights is more evidence that for many years someone failed to watch or speak up about the development that was going on at the lake, even allowing developers to sell land that the utility company owned. That land could have potentially buffered the water's edge and helped to maintain ecological conditions.

Green is the shoreline manager at the lake and he is also a familiar face in the community. His wife works with Tri-County Lodging Association, a group which is a tourism advocate and promoter for the lake area. He's gone to small lake towns to spread the word about the management plan and to try to ensure that residents receive correct information.

As the company continues to hash out a plan with the federal agency, Green is fielding questions from people at a Chamber breakfast who are scratching their heads on how it could be that someone who owns a house at the lake might have to surrender it because of someone else's oversight.

"We've got to come up with policies and do things differently in the future," Green says after a PowerPoint presentation that noted the timeline of shoreline-management plan events and how the situation evolved.

There are approximately twenty-seven thousand lake homes and only 6 percent will be affected, he reassures them.

In recent years, everyone has shown greater sensitivity about development around inland lakes, he says.

"At the end of the day, we will continue on the way we always have," he says.

But that still leaves some baffled.

"So you are saying that developers sold land that Ameren owned," says Craig Bischof of SCORE (Service Corps of Retired Executives), a small-business–nonprofit organization. "How can they do that?"

It's unlikely something like that can happen today, and everyone needs to move forward on how to correct the situation, Green says, attempting to move the conversation along.

One of the challenges is that some of the people are third- and fourth-generation owners who just came into the property and had nothing to do with its building or transfer of property transactions.

"At the end of the day, we need to do a better job of managing the shoreline," Green says.

Four months later, Ameren and homeowners breathed a sigh of relief. The Federal Energy Regulatory Commission approved an updated plan that saved the 1,500 homes in question. The houses stayed put, but the question of who exactly owned them remained and was supposed to be addressed in the future.

Rumblings of how development might somehow compromise the lake's natural setting began soon after the lake was formed in the early 1930s. Without proper zoning and planning, the lake's environmental problems were just a time bomb waiting to explode.

The first person to voice warnings of what development could mean to the lake was run out of town by promoters who wanted to capitalize on the prospective riches from waterside real estate and business. Ironically, he was in real estate himself.

Ward Gifford was president of the Kansas City Real Estate Board and began helping the newly created region promote itself. He also helped start the lake's first business association, the Lake of the Ozarks Improvement and Protective Association.

The organization formed in time for the Better Homes and Building Exposition of the Real Estate Board of Kansas City in February 1932.[4]

Businesses and towns from the lake created an exhibit with information for potential vacationers and homeowners.

In a March 15, 1932, article from the *Lake of the Ozarks News*, the association was described as being organized for a period of twenty-five years, to "assist in the education and training of the public by meetings, addresses, letters and publications in methods of improving the Lake of the Ozarks and regions surrounding it for the use and enjoyment of the general public and for the protection of its natural beauty and recreational advantages."[5]

Gifford served as the association's first president and other people composing the board came from across the state and lake area. He had established himself in Kansas City as a businessman who held jobs in a variety of fields, including working as a newspaper reporter.[6] He became known as a civic organizer who helped map out Kansas City's one-way streets downtown and headed an audit committee that kept watch on city hall.

In August of 1932, Gifford wrote a letter printed in the *Lake of the Ozarks News*. He cautioned readers to remain patient since the "boom which some predicted has not occurred." Rather, the slower progress is something to be thankful for, he contended in the letter. He asked for residents to hold steady and look for the tourists to come in the years ahead.

"If the influences which now predominate in the Lake of the Ozarks will by their own precept and example point the way to the proper sort of development of the lake, there is little question but those that come after them will find it necessary to build equally well or even better because public demand having been accustomed to the better things will not tolerate the cheap and the shoddy," Gifford wrote.[7]

His letter called for sanitary regulations, a road program, state parks, and game refuges.

"The laws now in mind for submission to the legislature have to do with the prevention of pollution of the lake waters, the protection of health, and the preservation of fish and game in the lake area."

Gifford soon unveiled a mission of the association, which included several progressive environmental stances. He wanted to protect fish and animals and work with all the state agencies to ensure a healthy environment, and he proposed "to extend an arm of protection around the entire Lake region."

So as the association began constructing roads that would accommodate the swelling population, it looked to create a scenic byway. Another idea was to create three parks, comprised of eighteen thousand acres, and then put them under public ownership.

These projects had hefty price tags, too. The parks were estimated at $1 million and the roadway would cost around $1.3 million, a great deal of money for the time.[8]

Further plans called for additional state park land to secure areas around the water so that fishing, boating, and recreation would be the lake's main forms of use.

Gifford also pushed for sanitation regulations and wanted restrictions when it came to home building and setting up subdivisions.

Protests began when news of the plans ahead for the area spread. Here was prime real estate next to water, with much potential for private development, and it would be protected from future building and moneymaking. Critics also scoffed at the costs of building the insular scenic byway and the public parks.

Gifford resigned under pressure in 1934 and did so graciously, saying that he planned to stay active in seeing the region prosper.[9]

His forward-thinking ideas for protection of the area were defeated by the other members' commercial interests.

The February 8, 1934, *Eldon Advertiser* ran a short article about Gifford's resignation.

"During his administration the Around-the-Lake road system, park planning, work for sanitation, prevention of stream pollution, the Ozark Trail, which is known as the inner belt road, and national publicity are a part of the things that have been promoted," the article said.

Lake historian Weaver thinks Gifford's ideas might have been too much for those wanting less preservation and more development.

"You have to read between the lines and know what was otherwise going on to realize his resignation was not particularly the reason stated," Weaver said. "Gifford was too progressive and too much an environmentalist for the members focused on development and tourism. After his resignation the association was reorganized, renamed, and new officers were elected."[10]

After Gifford resigned, the association moved its offices from Kansas City to the lake and eventually morphed into a tourism organization known as the Lake of the Ozarks Association.

Yet, it would be interesting to wonder what the lake would be like today if early Lake of the Ozarks founders had followed Gifford's lead, including his report about establishing healthy guidelines for preservation.

"Regulation of sanitary conditions already is under way, and this will be followed through next year," he had said. "Nothing should be permitted which will in any way menace the health of anyone in the Lake area whether he is a permanent resident . . . or merely on vacation."

One of development's consequences is people. They bring with them waste, whether in the form of sewage or in the form of polluted runoff from fertilized lawns or paved parking lots.

The state agency, the Missouri Department of Natural Resources, along with county health departments, have a long history of keeping track of failing sewer systems, municipal and commercial, along the shorelines, and one can infer from the records what has happened: Sewage and wastewater have been dumped into the lake.

Over the years, several studies have shown the compounding situations—outdated septic tanks, sewage spills, broken-down waste-water systems, and wastewater plants needing maintenance. A company called HTNB Corporation, an architecture and civil engineering con-sulting firm, conducted a study in 1999 for Lake Group Task Force that revealed up to fifteen thousand at-home sewer systems might be out of compliance and that many were in place before new standards took effect in the mid-1990s.[11]

But in recent years the warnings and sounds have grown louder and unraveled from unlikely sources.

In 2007, the Federal Energy Regulatory Commission mapped out guidelines for Ameren, the company that owns the lake and runs Bagnell Dam, to receive a new license. Among the requirements were to develop a biological conservation plan to ensure that the water could adequately sustain fish populations.

One requirement in particular has helped catalyze the modern water-quality movement at the lake. It was on page 82 of the federal agreement. Ameren would pay $15,000 annually for a five-year bacteria sampling program that the state Department of Natural Resources would conduct.[12]

The volunteer organization that seeks to protect the Lake of the Ozarks—the Lake of the Ozarks Watershed Alliance—called for people

Above: Old-time cars show an older image of the strip and the Lakeside Casino Restaurant, circa 1950s.
Below: Ameren, the utility company, has a prominent presence at the Bagnell Dam sign, as seen today. *(Photos courtesy of Missouri State Archives and Traci Angel.)*

Before and after photographs of Lowell's Boat Dock, at the west end of the Hwy. 54 Niangua Bridge over the Niangua arm of the lake west of Camdenton. Before: Lowell's Boat Dock, 1950s and 1960s. After: Present Day. *(Photos courtesy of H. Dwight Weaver.)*

These photographs show the contrasting styles of yesterday's boats to those of today. Before: Motorboat S.S. Dumplin. *(Photo courtesy of Missouri State Archives.)* After: Lake cruiser, Tropic Island, in summer of 2012. *(Photo courtesy of Traci Angel.)*

to assist with the sampling, and the Department of Natural Resources collected more samples than they could have using only their own employees. They tested for E. coli, a good indicator of fecal contamination because it is a kind of bacteria found in animals' feces and intestines. Each year during the program they selected fifteen coves to collect from over a five-month period.

In the spring of 2007, the group tried to drum up interest for volunteers. One of the differences with sampling this time around was the emphasis on bacteria in the coves rather than in the middle of the water body. However, no one was prepared for the community outcry when numbers were higher than expected.

E. coli is a bacteria found in humans and warm-blooded animals' intestines. As with other bacteria, dozens of strains exist, but many are harmless. It is, however, a helpful testing source. Its presence notes the probable presence of other pathogenic, or disease-causing, organisms, such as salmonella or hepatitis, said Scott Robinett of the Missouri Department of Natural Resources.

Often, higher numbers were recorded after thunderstorms because the rain could wash sediment containing animal or naturally occurring E. coli into the water.

Volunteers spent more than five hundred hours in 2007 collecting 396 samples in twenty-eight coves at 119 sites from Bagnell Dam to the toll bridge, the northern S-curve of the lake that includes heavily populated areas. Eight samples exceeded the 126 bacteria (colonies) per 100 ml geometric mean (the results can be found on a fact sheet on the department website under information about Lake of the Ozarks), which is the state water-quality standard for bacteria in recreational lakes. The Environmental Protection Agency's recommended maximum levels for bacteria in an individual sample is 235 (colonies) per 100 ml. While some of the samples were just a few dozen colonies over the approved limit, one in September of that year in the Jennings Branch Cove noted nearly two thousand colonies in the sample, indicating a problem. Another cove in Osage Beach tested in May showed nearly four times the state standard at 547 colonies per 100 ml.

Authorities decided to use a geometric mean to analyze the results. The mean is an average of test results for that reporting period.

"The state water-quality standard for E. coli is 126 bacteria [per] 100 ml as a geometric mean—a statistical method of comparing sets of data," Robinett said. "It is somewhat misleading to compare one sample result to a geometric-mean standard."

However, he said that Department of Natural Resources officials took every individual sample result that exceeded 126 seriously, inspected potential sources within the affected coves, and conducted repeat sampling. Repeated high numbers in areas such as Jennings Branch, and all results over 126, prompted inspections at the Department of Natural Resources' Springfield office.

Other problems were noted in the cove near Tan-Tar-A Resort, in which a residential area was the suspected culprit. Previously, the towns of Lake Ozark and Osage Beach had racked up violations for discharging sewage illegally.[13]

Detection of E. coli has to be taken in context, Robinett says. One E. coli in someone's well used for drinking water is a big deal, but one E. coli in a lake cove is nothing to cause alarm. E. coli is naturally occurring and could be linked to an animal or group of birds hanging around and defecating there. The E. coli count would be high if that happened, Robinett says.

The final conclusion of the five-year sampling in 2011 was that the lake was in overall good health because only forty-three of more than 1,600 samples exceeded the maximum of 235 per 100 ml, set as the standard.[14, 15]

However, when the conclusions came out, Natural Resources was still licking wounds from a scandal in 2009 that had the entire statehouse, lake business affiliates, and lake-goers on alert.

That was when reports of water samples that contained higher E. coli rates were released after the Memorial Day weekend, instead of a public notification before the weekend, putting vacationers' health at risk for related illnesses. The scandal revealed that the state agency decided to delay the release of information until after the weekend because of the fear it would keep tourists away.

When citizen watchdogs and Missouri governor Jay Nixon's Republican opponents caught wind of what happened, they demanded recourse and alerted the media. Headlines blared the news, which blamed the situation in equal parts on government's failure to look out for the people, heightened awareness of a potential problem at the lake, and political maelstrom.

There among the environmental controversy was Ken Midkiff of Columbia, Missouri.

He is known in Middle Missouri as something of a watchdog or muckraker. He has served as a Sierra Club lobbyist since 1993 in the Missouri statehouse. He's been appointed to environmental task forces and clean water and air committees.

In any given week, Midkiff is working on one Sierra Club cause or another.

He considers the Lake of the Ozarks a "sacrifice zone" for the people with money to have fun and leave the region's other waterways alone, but the withholding of information about high E. coli levels got his attention.

On learning of the delayed release of reports, he filed a complaint with state attorney general Chris Koster.

Midkiff said that agency violated the state's open records law because media and others had requested the information when they learned that Natural Resources had data that could pose a health threat to those swimming in the lake, but the agency did not act on the requests in a timely manner. "There's no question that they sat on [the information]," Midkiff says.[16]

Nixon, a Democrat, faced with embarrassment from an agency's perceived cover-up, declared he would intervene into the issue of improving the lake's water quality. He called for an all-out survey of the lake to assess the situation and an inspection sweep of all wastewater facilities with permits. The Missouri Department of Natural Resources' study revealed that 154 of 419 facilities had problems with permits to discharge water into the lake.[17]

The study also searched for E. coli, nitrogen, and phosphorus. They found high levels of E. coli at two of seventy-eight locations, levels of nitrogen that exceeded state standards at most locations, and all locations exceeded state standards for phosphorus.

Nixon said his administration would make the lake a priority, and the Department of Natural Resources continued to release information regularly about ongoing efforts there after the political fallout.

News about the lake and the Nixon administration's efforts to preserve it faded as time came closer to his reelection in November of 2012.

Other voices warn of what's ahead if something isn't done about sewer systems and regulations. This oversight could mean the pumping of sewage into the lake.

Bill McCaffree, an environmental attorney who has worked at the lake, attended a watershed meeting in 2010. His message contained a wake-up call.[18]

He told them that the debate is not whether septic systems are failing, but how to work for a comprehensive public sewer system that involves the lake's four counties.

Private systems and septic systems will have to be built, and it will take much money to build them. New properties and smaller communities need lateral sewer lines, he said.

"There's a great bit of difficulty and little cohesion in local government," he said.

He also mentioned the setbacks with regard to the splintered government entities.

"It's a development problem. Until something is done, it's a failure of entities to communicate," he says. "It's fractured governmental units with a lack of cohesion and government authority that regulates wastewater from the area."[19]

Another voice comes from longtime lake dweller, historian and author Dwight Weaver.

"The problem I see is that the chambers of commerce and developers and realtors have a biased approach and aren't going to tell you anything negative," Weaver says.

He has watched over the years as developers move in, strip the land, and then leave. He's recorded the transformation with his camera. Through his historical books he has documented a look back to the early days before the strip malls. When the builders go, what's left is a dense population only a short distance from the lake. Condominiums are stacked vertically, and sometimes, lack proper infrastructure.

"Everybody who has a condo, has a boat and a couple of the cars and they are paving so much of the area, what is the long-term effect?" he wonders. "They are replacing trees with buildings and paving over the ground."

Shopping centers have replaced the small novelty shops and their parking lots sprawl farther and farther into the outlying areas.

Crossing Bagnell Dam, multimillion dollar homes tower from shaved bluffs. It's a funny contrast to the family-owned barbecue joints, and remnants of old-time resort novelty shops dotting the nearby strip.

Weaver says a lot of people say there are too many condos.

"Development is going wild whether it's for the construction of condos or other kinds of buildings. There has been a big fight to keep condos from going above six stories but I believe that battle is about lost," he says. "They used to just build cottages on the hillsides and shoreline, but now they clear-cut and quarry away the hillside. They used to keep the trees for shade and natural beauty," he laments. "Now they don't."

Then comes the conflict of getting people involved: many live there part-time and don't have a fully vested interest in what happens.

They come on Thursday, Friday afternoon. They don sarongs and Peckers restaurant tank tops, tropical flower-splashed swim shorts and ratty sun visors. Bass boats, speed boats, cigar boats, and wave runners launch into murky water gaining speed as they decrease in size fading from the shore. Water wakes climb as the sun glides higher.

"The majority of our population is weekenders. If they are on a sewer system [in Saint Louis or Kansas City], they have no idea how to operate [a single-home septic system]," says Donna Swall, executive director of the Lake of the Ozarks Watershed Alliance.[20]

The Osage River water turns over annually, so what difference does water quality make since pollution won't stick around?

Jennifer Byers says sewage that leaked near her home in 2005 disrupted her family's life. Their dog Sammy's fur was covered with fungus. His hair was falling out.

"I had a phobia about my kids swimming," she says. "My concern was that [the water] could get worse."[21]

Notices of violations were filed by the Missouri Department of Natural Resources in June and July of 2005 stating that the sewer main the city of Osage Beach operated near Byers's cove off state highway KK was malfunctioning, sending sewage into waterways. Rick King, Osage Beach superintendent, said the underground pipe had broken and a crew repaired it immediately.[22]

When Byers's kids started having diarrhea, she took samples to a water-quality lab. She says the results indicated *E. coli*. Her concern heightened because her infant son's baby formula was made with the contaminated water. She pursued the problem with city officials and the Department of Natural Resources. She grew frustrated.

Then she gave up.

"The damage was already done and the only thing that I was concerned about was my infant son's health, and he was okay. We shocked the well [a chlorination process to treat the problem], and moved out five months later," she said.

Finding fecal matter in a water sample might just indicate there's been a warm-blooded animal nearby, said Paulette Mitchell at H20 Lab, the private water-quality-monitoring business in the lake area where Byers took her sample.

You can't blame water issues on the swelling populations or keep people from moving to the lake, Mitchell said.[23]

"You can't open up paradise and then close the door."

A strong contingency is making sure to keep the door open wide and ensure the lake's positive image.

The Camden Chamber of Commerce and many businesses are part of another group called Citizens for the Preservation of the Lake of the Ozarks. It propagates marketing materials to combat what the group views as unfair media attention to a few high-bacteria counts and the post-Memorial Day political scandal that followed the Department of Natural Resources' delayed release in 2009 of high E. coli reports until after the holiday weekend.

The group was headed by business leader Greg Gagnon, president of a local bank, who died unexpectedly in the fall of 2012. But Jim Divincen, executive vice president of the lake's Tri-County Lodging, an organization that promotes area businesses, has helped educate the public about the group's work.

In 2009 when news came out about the E. coli at some of the lake's public beaches, most of Divincen's colleagues began scrambling to clean a now-tarnished image.

Divincen described it as being caught in a political football game. Heavy rain added to the high numbers of recorded E. coli, as rain runoff adds higher E. coli to any body of water, he said. The delayed release of information made it worse.

"It created a press feeding frenzy," he said.[24]

Republican state lawmakers Brad Lager and Kurt Schaefer cried foul against Democratic governor Nixon and his state agency. Then Nixon called a press conference at the lake, and said the water was unacceptable. Some people caught up in the controversy lost their jobs, and the Department of Natural Resources released a water-quality report later in December 2009 that showed the lake's numbers in the overall healthy range. The governor declared the lake safe.

"What changed?" Divincen asks.

Divincen and the Citizens for the Preservation of the Lake of the Ozarks have studied up on their science so Divincen can rattle off statistics from a PowerPoint presentation that he takes to meetings. The presentation is his attempt to educate the public about the situation by showing everything from the history of the E. coli water sampling to the scientific details of bacterial strains.

"We're getting beat up at the state parks," he said. "In the public's mind when the beaches are closed, the lake is closed." They cancel reservations and avoid making the trip.

So Divincen and the preservation group began backing further studies, including some involving the US Geological Survey and state Department of Natural Resources to test beaches.

The group set up a website, www.LakeWaterQuality.org.

The website is a public-relations attempt to refute rumors about bacteria or pollution in the water. The Department of Natural Resources reports and quotes from experts, along with links to national magazine articles that put the lake on top healthy lists, are among the information for readers.

This group is different than the other watershed association because it focuses on businesses and the short-term, Divincen explained.

Divincen said he supports policies which mandate that sellers must adhere to new regulations and repair outdated sewer systems before selling properties. And he's adopted a philosophy that "all scientific information should be reviewed when evaluating any body of water," Divincen said.

CHAPTER 4

Those Who Built on Ameren's Land

Perhaps no other issue more clearly demonstrates the example of what happens when heads look away from development than when the Federal Energy Regulatory Commission (FERC) asked Ameren for a shoreline-management plan, as part of its relicensing agreement in 2007.

The federal agency required the shoreline-management plan before it would grant Ameren another forty-year license for the hydroelectric operation.

The plan's purpose was to ensure that a buffer was created to preserve the environment around the lake, as the agency mandates, and something that is required for the utility to operate the dam at the lake.

But a proposal to put the boundary elevation at 662 feet above sea level opened a can of worms that dated back to 1937.

During the relicensing process, it was discovered that real estate agents had sold land to people for lake homes that was within Ameren's boundary.

At first, the federal agency said all the structures within that boundary would have to be removed.

Homeowners, like Peggy Crockett of Gravois Mills, were livid when they learned they didn't own the property where their houses sat.

Deeds were filed after the dam was completed in the early 1930s, and the utility company at the time, Union Electric, shifted property ownership to Bank of New York Mellon, which became the indentured trustee after investors bought bonds to finance the project. This was done as collateral for debt the company owed.[1]

Somehow, though, some 1,500 homeowners managed to build homes in the company's property around the lake.

When Crockett and others found that they were essentially trespassing on land they didn't own, even though they had deeds for the property, they started making noise.

Crockett and her husband cleared the land and built their home more than fifty years ago.

"We all own to 660 [feet above sea level] and have deeds and title insurance to that fact," Crockett wrote in a public response to Ameren. "I am 84 [in 2012] years old and you have upset people with your trying to take our property when the lake is all full of property owners. It is now close to 1 million people here. We bought our property from Union Electric. If we had not bought [land], and helped to maintain the Lake, it would be quite different now. Please leave us all at 660."[2]

Others felt that sentiment and Crockett remained upset even after the federal agency eventually accepted an agreement from the utility company and made allowances for those structures within the boundary, but set the boundary at 662 around much of the lake, with exceptions to draw the boundary around those homes built lower.

"They still think they can take two foot of our property and that's not right," she said a few months after the decision.[3]

Crockett heard about lawsuits against Ameren and federal legislation to help those caught in the reinforcing of the original property lines.

Crockett and her husband, who had died fifteen years earlier, got into the area before the boom and, like others, they profited from it. They settled there in 1958 and lived there permanently beginning in 1977. They helped to develop a subdivision off the Osage River in 1960 that created twenty-five homes called Maple Hill.

She and all the property owners can trace the plats back to 1930, where the files in Warsaw, the Benton County seat, show that each have a right to the shoreline, she says.

"We all came here to retire and spent all our retirement in maintaining our homes, and now they are saying we don't own it even though we have deeds," she said.

But how could something like this happen? Wasn't someone watching?

Not really, said Warren Witt, hydro-operations manager at Ameren who handled damage control at the time, much of it to do with the boundary issue.[4]

The problem began shortly after the lake formed.

When the lake was built, Ameren controlled thousands of acres of land, Witt said. A federal law, a public utility holding act in 1935, mandated that utility companies be subjected to some regulation. One of the law's purposes was to keep utility businesses from participating in business ventures, such as real estate, unrelated to their intended energy purpose or business.[5]

Among its effects on Union Electric was that the company began to sell off some of the land surrounding the lake, Witt said.

Where the trouble began was when developers started creeping into Ameren-owned land and selling it, according to Witt.

"They crossed the property line, and a developer on accident, or purpose, gave someone a deed to Ameren-owned property," Witt said.

They sold an incorrect deed that the title companies or a surveyor should have caught.

Witt pointed to a map and noted the elevation changes.

Ameren is innocent, he said.

"We were running a power plant," he said. Houses were built, and Ameren wasn't always looking.

"No one knew, and no one was watching," he said. "Frankly, we didn't care. Prior to the environmental regulations in the 1970s and 1980s, we didn't need the land."

The environmental movement in the 1970s spurred more regulations to keep the land and water around the lake clean and protected.

Before 1980, the Army Corps of Engineers helped regulate the shoreline, but then the corps' responsibilities shifted away from enforcement, Witt said. The utility company developed a shoreline-management office to help with the shoreline permits.

If today's technologies had been around then, they would have helped surveyors and others accurately determine property lines, Witt said.

In the end, the boundaries were redrawn to 662 elevation everywhere except where there is a house below that elevation, then the boundary goes lower to preclude the house from being inside the boundary.[6, 7, 8, 9]

The fight then became for Ameren to relinquish property rights to those homeowners. Ameren was expected to "quit claim" the property.

"We have filed with each of the four counties for an estoppel certificate [a legal document needed for property transfer] if the house is on the property and are not going to require them to tear it down." Witt said.

Confusion over who owns the land won't happen again because the company has several people on staff, and uses technology and aerial photographs, to keep track of the situation, Witt said.

If this lake were built today Ameren would have kept the land, Witt said.

Other comments from the boundary quandary echoed the longstanding developmental challenges.

"It is long past time to hold the counties, the builders and developers, the title companies, and the Realtors accountable for decades of errors," said Lita Francis of Osage Beach in her public comments to Ameren. "This boundary mess is the result of years of unbridled, uncontrolled development with no regard for boundaries which have been in existence for eighty years. There is a pervasive lack of knowledge and regard for property lines in this region. It happens in my area, too, although I don't live in this disputed area. FERC (Federal Energy Regulatory Commission) has every right to do its job since the main purpose for this dam is to generate energy."

Francis went on to say that she felt Ameren's plan was generous, and that the lawmakers making noise about the federal agency's relicensing process were playing politics.

"I urge you to more closely monitor further development along this beautiful lake shoreline. I realize real estate is not your function; but there needs to be a mechanism in place to prevent these deed and plat errors in the future," Francis wrote.

As tempers rose, many people pointed fingers at the federal agency. While doing damage control, the commission released information about how Ameren's primary responsibility was to "implement the terms of its license."

Under its previous and current license, Ameren was supposed to "prevent the construction of unauthorized structures inside the project boundary, otherwise known as encroachments, and to take appropriate action to ensure that neither project purposes nor the expectations of

Before and after photos of Mark A. McBride's family cove, taken from the north shore of Mill Hollow Cove, where the Niangua and Osage Rivers meet. One is taken in 1976 and the other in 2010. The photos show the contrast in development of the shoreline during the time lapse. *(Photos courtesy of Mark A. McBride.)*

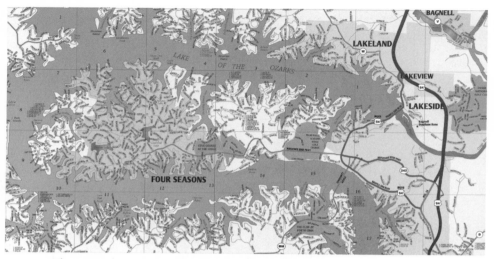

This 2012 Lake Media Lake of the Ozarks Community map shows the numerous street names and development of the many coves around the lake's popular channels. (*Courtesy of Lake Media.*)

the structure owners were unduly affected," commission officials said in 2011 press release, an effort to set the record straight.

"Over many years, Ameren failed to carry out this obligation. Ameren's repeated failure to properly implement the terms of its license has allowed matters to get to the point where it does not even know exactly what structures have been built within the project boundary and whether they were authorized. FERC recognizes that Ameren's failures have left local property owners in an extremely difficult position," the press release stated.[10]

CHAPTER 5

Impaired Waters—Yes, There's Evidence

*In wine there is wisdom, in beer there is Freedom, in water
there is bacteria.*
                         —BENJAMIN FRANKLIN

On a mild fifty-degree day in January 1932, the Reverend Gentry Patrick
led parishioners from his Baptist church in Roach, Missouri, to the cool
wintry waters of the Niangua River. The group, dressed in their Sunday
best, strode toward the docks of Niangua Bridge Camp that kissed the
Lake of the Ozarks' tributary. The group descended into the caramel-
colored current and emerged reborn through the Christian tradition
of baptism. It was the first ceremony of many held after the lake's
settlement.[1, 2]

Eighty years later, the Environmental Protection Agency had designated three places as impaired in that same Niangua watershed because
of poor water quality that threaten plant and animal life.

The federal agency assigns the listing of "impaired" to bodies of
water that fail to meet water-quality standards the agency has set. These
water systems are considered too polluted for their intended purposes.[3]

One section of the upper Niangua is highlighted because of high
E. coli, bacterium commonly found in fecal material. Another section
was noted for high amounts of phosphorus, high enough for algae to
consume and smother other lake areas since algae can take over and
push out other natural-occurring ecosystems.

Elevated recordings of both *E. coli* and phosphorus, while occurring
naturally in some quantities, is evidence that the lake has changed over
the years. But both can also be the effects of the dumping of sewage into

the waterways from surrounding sewer systems, septic tanks, and natural runoff from farmlands upstream that could have wreaked havoc on the lake and its offshoots created from the damming of the Osage River. Nitrogen and phosphorus are key nutrients found in sewage and wastewater and can also mean eutrophication, an unhealthy condition that negatively affects aquatic life.[4]

Despite what lake dwellers and advocates want to say and believe, there is evidence that influences from development and human influence are harming the lake.

The signs come in the forms of actual water quality, such as in the readings of bacteria within the water, or high concentrations of nitrogen and phosphorus. And it comes in the form of dying or threatened plant and animal life due to the impact of development. Species that once thrived are fighting for habitat space. They are marked as endangered and their days are numbered. Other living things have become invaders, sucking out the native plants and animals.

Protecting the Water

In 1972, the US Environmental Protection Agency made amendments to what is known as the Clean Water Act. The actions put guidelines in place for discharging pollutants from point sources (pipes or linear points of discharges) and gave the agency the ability to set limits on pollutant levels to ensure the water quality was maintained. Those who want to discharge pollutants must obtain a permit first. The EPA has passed authority of the permit program to state governments, but still retains oversight when needed.

Homes on municipal systems and those that use septic tanks do not need a permit because they operate under an assumption that they are responsible in handling waste. However, industries and facilities do need a permit if discharges go directly to surface waters, the EPA states.

Missouri's Water Pollution Control branch, a division of the Natural Resources' Water Protection Program, issues permits for point sources of water pollution. The majority are for treated wastewater from domestic and industrial facilities and have five-year cycles.

A more modern push for water quality came from the EPA in 2001 when the agency called for nutrient (nitrogen and phosphorus) criteria to be a starting point for states. The agency recognized that too many or too few of these nutrients could pose harmful effects on a water body. For example, excess "nutrient loading" could mean rapid growth of algal blooms and depletion of dissolved oxygen.

Missouri developed a plan for addressing these requirements in 2005, but was still working on putting them in place eight years later.[5]

Officials with the Department of Natural Resources explained that the state agency did adopt criteria for lakes and reservoirs in 2009, but the EPA denied approval of part of it. The two departments were still working on revisions in 2013.

Non-bacteria Pollution

Nitrogen found in sources, such as fertilizers, manure, and wastewater from septic systems can combine and affect surface water and can contribute to pollution.

Scientific studies have documented that nitrates in freshwaters and coastal waters could cause eutrophication and the growth of algal blooms, which can be detrimental to aquatic ecosystems. Phosphorus, too, can be in surface runoff and wastewater discharge. Too much of this phosphorus can stimulate the growth of undesirable organisms and cause problems associated with eutrophication, which is defined as "an increase in the nutrient status of natural waters that causes accelerated growth of algae or water plants, depletion of oxygen, increased turbidity, and a general degradation of water quality," according to Gary M. Pierzynski et al. in *Soils and Environmental Quality*, a soil science textbook.

With eutrophication comes negative effects on water life. Excessive weed growth can lead to the depletion of oxygen during plant and algal decomposition. This oxygen is needed to maintain a healthy fish and aquatic life population.

It was higher levels of the nutrients nitrogen and phosphorus, which were above both federal and state guidelines, that put the Niangua and Osage arms of the lake on the EPA's impairment list.

After the EPA rejected the state Department of Natural Resources' proposal to establish nutrient criteria guidelines, the federal agency gave this response:

> Section 7(a)(2) of the Endangered Species Act (16 U.S.C. § 1536) requires that federal agencies, in consultation with the United States Fish and Wildlife Service, ensure that their actions are not

likely to jeopardize the continued existence of federally listed threatened or endangered species or result in the destruction or adverse modification of designated critical habitat of such species. Regarding today's approval actions, the EPA is making its decision subject to the outcome of consultation under section 7 of the Endangered Species Act.[6]

Pollution consequences can mean health threats to both humans and wildlife.

The Missouri Department of Health issued a chlordane warning in area paddlefish in the late 1990s. Chlordane is a known carcinogen and had been detected in paddlefish to the extent that the department advised against consuming the fish in 1990. The chemical was banned in 1988 and seven years later the detection of it in the paddlefish was low enough that the warning eventually was lifted and paddlefish were considered safe to eat.[7]

Bacterial Problems

University of Missouri graduate student Rebecca O'Hearn spent two years in 2007 and 2008 collecting samples around the Lake of the Ozarks to conclude that the area's urban development and use of septic tanks in the city of Lake Ozark soils could mean pollution into the lake's swimming coves.

For her thesis, she spent the summers testing for nutrients E. coli and Bacteroides thetaiotaomicron, a bacterium found in the human gut. By testing for this particular bacterium, O'Hearn's numbers could perhaps pinpoint which readings came from human influences, such as leaking sewage.[8]

O'Hearn expanded on research from the 1980s that Jeffrey D. Mitzelfelt conducted. His work noted the effects of on-site wastewater treatment on water quality. His thesis noted specifically human influence on water quality when he studied twenty-nine coves in three channels over the course of three years. In his report, he found that mean fecal coliform concentrations were fifty times higher in coves with development than in the lake's main channel. How much the coves exceeded the standards was proportional to the amount of building and development around them, his thesis said.[9]

His thesis concluded that current sewage-treatment methods were ineffective in removing pollutants affiliated with wastewater.

Since Mitzelfelt's research, some of the bigger towns like Lake Ozark, Camdenton and Osage Beach have installed centralized sewer systems. Some smaller ones, such as Gravois Mills and Laurie, were working on creating theirs.

Much had changed by the time O'Hearn started her research. Cove housing units had increased by 40 percent. And, as O'Hearn's thesis notes, wastewater-treatment facilities received permits to discharge effluent, sewage, and other treated wastewater, into the lake. Plus, many residences still depended on on-site sewage systems perhaps without any updates.

O'Hearn's results did indicate that human gut bacteria were found in the lake, as her study mentioned, but she concluded that rain events in the coves where she tested perhaps masked some of the results. Bacteria from soil runoff potentially skewed results for *E. coli*, she said. O'Hearn also noted that hydrologic influences, or how the water flows, from the main channel often made it difficult to analyze the cove.

An overall conclusion found checking for *Bacteroides* was a better indicator for human influence than just *E. coli*.

After her findings, O'Hearn concluded that the lake's health overall was good, but some of the coves with lower flushing rates, or the time it takes for the water to flush out, experience *E. coli* spikes during rain events from sediment and septic system runoff or sediment resuspension.

"It depends where you are," O'Hearn said. "The lake is a huge reservoir, and there is a lot of water in the main channel to dilute pollution, whereas in the coves there is less water to dilute pollution."[10]

Even if pollution is diluted to accepted levels, whether pollution in general is okay is left for society at large to judge, she said.

But one of O'Hearn's findings did note a problem that would later be identified as a concern.

A cove in the Grand Glaize arm located in the state park had the highest readings for *E. coli* and *Bacteroides*. This area also contains a lagoon that treats wastewater from park visitors.

Two years after O'Hearn's study, the US Geological Survey, Missouri University of Science and Technology, and the Missouri Department of Natural Resources teamed up on a project that sampled near the Grand Glaize Beach cove and the Lake of the Ozarks State Park.[11]

Preliminary data that Natural Resources posted showed *E. coli* counts as high as 5,300 colony-forming units per 100 ml of water. To put this number into context, the limit used to determine whether to close public beaches is 235 per 100 ml.

But how bad is the water quality of the lake? Does the EPA's rating give an accurately placed scarlet letter?

Science moves slowly and while there are recordings of higher numbers of bacteria, those recordings can quickly fall by the wayside, as they are quieted with information and data that show otherwise.

One of the more comprehensive studies targeted coves where treated wastewater was discharged.

Concerned officials for the city of Osage Beach brought in Springfield-based analytical and testing firm, Consulting Analytical Services International Inc. in the early 1980s to check out any sewage discharge in the lake and the impact it had on water quality.

One of the purposes, like much of the other previous and current research, was to see how development and human populations affected the watershed.[12]

For the analysis they targeted developed, intermediate, and undeveloped coves for a rounded assessment in the Turkey Bend/Osage Beach portion of the lake.

Other samples were taken ten to twenty yards from discharge points, where treated wastewater systems were allowed to dump.

The most significant finding was in the coliform bacteria levels. The study showed 43 percent of samples exceeded the limit set for state whole body contact areas in water. The samples were taken during the recreation season from April 1 to October 31, 1981.

The septic leachate analysis and other water samples noted forty-five leachate plumes, with twenty-three from residential, fifteen from resorts, and seven of unknown sources. While large numbers of the plumes were attributed to rainfall, others might have been undetected because of dilution and water levels, the report said.

"Results of water chemistry analysis support leachate shoreline scanning results and reflect an increase of nutrients (total phosphorus and nitrogen), chlorides, and fecal coliforms in septic leachate plumes. Chemistry results indicate that growing ecologic, aesthetic, and health problems can be expected if the observed septic discharge continues," the report said.

Included in a 1974 study geared to provide a plan for the lake's watershed, an EPA report said, "The water quality is good. A few locations showed degradation as a by-product of human activity during the high use periods, especially on weekends and during summer months. While such degradation was readily detectable, there was no indication the Lake was seriously polluted or contaminated." However, a few pages later in a section titled "Existing Economic Base" it was noted that "the potential for contamination of groundwater supplies increases with each new well drilled and each septic tank system installed. The Missouri State Division of Health and other authorities believe a dangerous stage has already been reached."[13]

Tony Thorpe, codirector for the Lakes of Missouri Volunteer Program, has spent the last decade training volunteers to collect samples from waterways across the state. The program was created in 1992 to track four lakes in the Kansas City area. Over the years, volunteers have collected samples at dozens of Missouri lakes. The program's goals are to monitor the water bodies over time and to educate the public.

Thorpe was finishing a master's degree in fisheries and wildlife with Dr. Jack Jones of the University of Missouri when he coordinated the program, funded 100 percent through grants.

Because the program focuses on water quality pertaining to nutrients, rather than bacteria, volunteers have mounds of data analyzing nitrogen, phosphorus, and chlorophyll levels.

When state officials and other volunteers started testing for bacteria and E. coli, Thorpe cringed. He thought the numbers would be high.

But the water is not that bad, Thorpe said.[14]

"It's funny, the water quality is no worse than twenty or thirty years ago, and it's actually improved," he said. "The numbers are way lower than expected."

The lake has a couple of hot spots that could probably be cleared up with direct efforts that monitoring might reveal.

Septic tanks might contribute, but the runoff from concrete and the sediment that accompanies the runoff can also affect the water quality.

Growth slowed at the lake during the economic downturn in the

City of Lake Ozark received a notice of violation from the Missouri Department of Natural Resources in 2007 because of untreated wastewater illegally discharged into the lake. In September of that year, according to an EPA news release, state officials noted that a lift station couldn't handle the effluent load and sent ten thousand to fifteen thousand gallons of raw sewage into the lake. The Osage Beach sewer system also had numerous overflows from stations near Grand Glaize Bridge that began in

1995. The system was unable to hold the high-volume vacation sewage loads until upgrades were later made (Stoner, 2000). This series of photos from September 2007 shows a lift station overflow (above) and the septage flowing near Jennings Branch Cove, Lake of the Ozarks. *(Photos courtesy of Troy Potteiger, The Missouri Department of Natural Resources.)*

early 2010s, but if the population picks up again before changes are in place there could be a turning point for the lake to go from great to crappy, Thorpe says.

In 2011, Thorpe and the University of Missouri–sponsored program selected the Niangua arm for a "snapshot sample." Their original idea was to monitor the entire watershed over four days, but realized the expansive nature of the project and instead focused on the affected Niangua watershed with the EPA's concerns in mind.

The data they collected revealed little about pollution sources.

"We didn't get enough information to tease it out because it was a really weird high-flow period," Thorpe says.[15]

Some of the samples did have high nutrient contents, such as nitrogen and phosphorus, but the group couldn't pinpoint a source such as agricultural runoff, wastewater pollution, or runoff from development and pavement.

In the case of the Niangua, the sampling volunteers concluded that to address water quality, one needs to "look beyond the shoreline."

Efforts to improve quality, "shouldn't be restricted to managing the shoreline," Thorpe said. Instead, the focus has to broaden across the wide Lake of the Ozarks watershed.

That means every tributary and every adjacent property, where pollutants lurk and could leach into the soil and ultimately cloud water sources, is a factor when it comes to creating a healthy lake.

Theories can abound and finger pointing allows for many conclusions without clear evidence as to why the lake is polluted.

Publically-filed pollution complaints to the EPA for this region are small.

The EPA provides more of an oversight role and receives leads through a citizen-driven website. The agency also steps in when the state's Department of Natural Resources officials request help. EPA's funding is funneled through these state departments, said EPA Region 7 spokesman Kris Lancaster.[16]

A preliminary search from the EPA enforcement division brought only about a handful of complaints in the Lake of the Ozarks region for fiscal year 2011. However, this number may not be comprehensive because there are several different departments reporting, and some complaints could be filed under different departments, Lancaster said.

To put the water quality in perspective, the regional agency counted 140 complaints in Missouri with five of them in the Lake of the Ozarks. Another sixty-seven complaints were filed in Iowa, Kansas, and Nebraska for that same time period.

The Species Lost and the Unwanted Species Gained

One of the true markers of a species lost is the Niangua darter. Federally listed as "threatened" and state-listed as "endangered," these fish historically would have been found in those waters, but now find it difficult to survive. Their habitat, which is small to medium rivers, is gone with the lake's enclosed reservoir and stream channelization. No longer to roam the rivers, they are stuck to adapt in the nonmoving lake waters.

STATE OR FEDERAL LISTED ENDANGERED ANIMAL SPECIES WITHIN NIAGUA WATERSHED

COMMON NAME	FEDERAL STATUS	STATE STATUS
Lake sturgeon	—	endangered
Niangua darter	threatened	endangered
Gray bat	endangered	endangered
Indiana bat	endangered	endangered
Black-tailed jack rabbit	—	endangered

Source: Missouri Department of Conservation

The darter was among eleven fish within the Niangua watershed classified as rare or endangered, and state conservationists have kept a close eye on it since the 1980s.[17, 18]

Another indicator that the ecological landscape is changing is not the disappearing species, but the unwelcomed ones. The most significant modern nuisance is zebra mussels. The exotic invertebrate can push out native mussels and feast on food sources, such as plankton.

Greg Stoner has worked with the Missouri Department of Conservation since 1991. He has offered talks about invasive species, such as zebra mussels, in water bodies, including the Lake of the Ozarks. Zebra mussels likely came to the lake via boats, attached to the hull and inside the motor housings and wells. When the mussels feed, they filter microscopic plants and animals, the same food vital for bass, shad, and small fish.[19]

They have different impacts, depending on characteristics of the lake, Stoner said. They can be an annoyance, or they can create real problems if the numbers are enough.

"When they finally get established all through the lake, we don't know what it means to existing fish populations," he said. And all reports say they are spreading, he said.

Zebra mussels also provide a natural water filter, which can mean more sun for lake plants. As the plants grow they can cause problems for boats and swimmers in shallow water, according to information from the Lakes of Missouri Volunteer Program. Zebra mussels could change the lake's natural habitats for plants and aquatic animals.

Another exotic species, the Chinese mystery snail, was first reported in 2008 and has been found in the Niangua River. It is known for its ability to multiply rapidly and squeeze out other natural-occurring species as it feeds and propagates.[20]

Harm to Fish Populations

To sportsmen and sportswomen, the Lake of the Ozarks, as is much of the Missouri-Arkansas area, is known for its fishing.

Hundreds of tournaments are held there annually, and the US Census Bureau's retail sales for the counties of Camden, Miller, and Morgan put taxable sales, with much coming from the fishing industry, at a value of more than $1 billion annually in recent years.[21]

The Niangua Watershed has recorded nearly a hundred species, and the varied habitats make it appealing for a variety of fish populations. Bass, trout, paddlefish, walleye, catfish, and crappie are just some of the species drawing anglers to the waters.

But the area has had a history of fish kills related to the Truman Dam in the upper watershed and the Bagnell Dam that created the Lake of the Ozarks.

The Missouri Department of Natural Resources noted that between 1978 and 1994, the lake was on the EPA's list of impaired waters because of chronic fish kills that occurred due to turbulent flow from the dam. Earlier reports also indicated low dissolved oxygen and gas supersaturation, a condition that occurs when dissolved gases become greater than the atmosphere, and can create circulatory and other health problems for fish.[22]

Even when conditions in the water calmed, the lake remained impaired because of fish trauma when they were caught in water flowing over the dam during flood control or when power was generating. Other times fish was caught on the screens and barriers meant to keep fish away from the dam's hydroelectric equipment.

Some fish species, such as paddlefish, are stocked annually because Truman Dam blocked fishes' spawning habitat at the lake.

In 2003, Ameren, which operates Bagnell Dam, released a report called the Lake of the Ozarks Historical Fishery Data Summary. In it, the company documented fish habitat stocking information and other trends about maintaining fish populations.

One group, the Missouri Chapter of the American Fisheries Society, decided to speak out against the document, and its officials said it did not thoroughly and accurately represent what was going on with the fish kills at the lake. The chapter's president wrote a letter to the Federal Energy Regulatory Commission after the report was drafted.

The letter notes that while the authors of the report provided much background about fish species and their habitats, it fell short in including historical documentations of fish kills over the years.

"The most glaring of these is the complete omission of historical accounts of the many documented fish kills at the Lake, specifically those associated with the project operation itself. Since the draft report is dated September 2002, we see no reason that it should not include documentation of at least the significant paddlefish kills that occurred in the autumn of 2001 and spring of 2002," the letter states.[23]

The fisheries society asked that not only that these kills be documented, but that Ameren include what it was going to do to correct the situation. The group also questioned the scientific merit of some of the information provided.

"It was disturbing to note the lack of peer-reviewed technical literature used (outside of a limited number of Missouri Department of Conservation technical reports) to substantiate claims and conclusions.

This draft report is being prepared for use in an important process that will affect important aquatic resources for many years. Thus, the report should conform to the careful scrutiny typically accorded a scientific publication. In its present form, this draft report fails that test," the letter stated.

A few years later, in 2005, a judge ordered Ameren to pay $1.3 million for fish kills recorded in 2002. The Missouri Attorney General's office claimed that nearly forty-three thousand fish were sucked into the dam's turbines in May and June of 2002. Governor Nixon had filed a lawsuit against the utility in 2003 on behalf of the Clean Water Commission. The court order placed $1 million in the Conservation Commission Fund and the rest in a Natural Resources Protective Fund.[24]

In addition to the money, Ameren was ordered to construct a net barrier to protect the fish from harm. The agreement also ordered the company to: pay $134,000 annually to the Missouri Department of Conservation for fish stocking; pay $2.1 million to the Missouri Division of State Parks and Historic Sites to mitigate the impact of erosion, install more efficient turbines, and require the dam to meet water quality standards for dissolved oxygen levels; and pay $175,000 annually for activities under the Endangered Species Act.

It took a few years for the ruling to make an impact.

Ameren's 2011 Corporate Responsibility Report boasts about protecting the Lake of the Ozarks' fish population, but fails to mention why the company took the measures.[25]

"We've taken steps to protect aquatic life" is the heading of one section.

"Lake of the Ozarks, Ameren Missouri's hydroelectric energy center, installed 900 feet of netting across a portion of Bagnell Dam to act as a barrier and keep fish from running into turbines at the facility," the report stated. "Since 2009, the fish net has been successful in protecting fish at the lake, reducing fish mortality significantly."

CHAPTER 6

The Land and Geological Factors

To a person who is not a scientist, a lake might appear as nothing more than a large puddle of water. Under the wrong conditions, and with enough pollution, it could seem to take the form of a lagoon or a cesspool.

The truth is much different. A lake's formation, its years of geology, and its recorded limnology—or the study of bodies of freshwater—reveal that a lake is actually a dynamic phenomenon or creature.

"Limnological characteristics of large reservoirs differ to some extent from those of natural lakes as a result of human input in the form of regulation of water levels for flood control and power generation," states an environmental assessment of the Lake of the Ozarks prepared in 1982 by the US Army Corp of Engineers.[1]

It also was noted in the assessment that "lowest pools usually occur during winter. Filling generally occurs during the spring months via precipitative runoff from the watershed. During summer, water may be withdrawn for power generation or for flood control."

Beneath the darkness lurks a world dependent on a symbiotic relationship between the water and its contents, as well as the living plants and animals that call it home. The structure of lakes can be described as layered, such as a forest ecosystem. A lake that is a reservoir has other variables by way of damming; its health and water quality are at the mercy of the officials who control the release of water.

A variety of life-dependent components are needed to keep a lake thriving. Among them are light and a lake's watershed health, including what development is around the body that might influence erosion and pollution.[2]

As for light, the further the sun penetrates through the water's depths the more plant-life processes called photosynthesis can occur, and thus the more life the water can sustain whether in the form of algae or aquatic plants.

A lake's watershed often holds the ingredients for its health: Good stuff, such as dissolved oxygen from the atmosphere and plant life native to the area, along with the bad, such as sediment from erosion, nutrient pollution from fertilizers, and harmful and improperly managed waste-water spilling from parking lots and pavement. Rainwater that drains down from surrounding land into the lake can bring with it other concerns, such as the *E. coli*–tainted soil that is washed into the Lake of the Ozarks after heavy rains. This runoff affects what is known as the lake's chemistry. Pollution can mean heavy metals or certain toxins are flushed into the waterways and can create havoc for the living things there.[3, 4, 5]

Even getting past the disgusting thought of sewage and potentially harmful bacteria that accompanies it, there can be problems with the nutrients the sewage contains.

Sewage contains a much higher concentration of phosphorus, which can prompt algal growth. Density of the algal blooms suffocates other plants and fish and causes changes in the aquatic ecosystems.

A lake's geology can be one of the biggest indicators of its health and future water quality. It is affected by everything that surrounds it. Any addition of outside sources in the form of pollution or sewage can change its chemical and physical makeup.

For the Lake of the Ozarks, the topography makes it easier for the lake to be polluted. That topographical culprit along the lake's water-shed is known as karst.

This type of landscape accounts for 10 percent of the earth's surface and is distinctive because the rock that forms the landscape (mostly dolomite, which is like limestone) can dissolve as the surface water meets the rock. Cracks grow larger. Crevices expand. This process of breaking down rocks, usually limestone or dolomite, accounts for the many sink-holes and caves that are common within the Ozarks. The sinkholes take on the role of a funnel and deposit the water directly into the area's water supply.[6, 7]

This characteristic, though, also means that the ground and rock

A sinkhole at Hahatonka Spring had become a dumping ground. *(Photo courtesy of the Missouri Department of Natural Resources.)*

loses its natural filtering dynamic and that any polluted surface water flowing over the karst area can easily penetrate the soil.

The groundwater is replenished or recharged in two different ways. One is through rain and other forms of precipitation, which gradually finds its way to a spring or water source. The second way is from another area in the watershed. Groundwater can travel long distances without the natural filter of soil and organic material, and the surface water and groundwater interchange freely. Often, a larger amount of water can find its way into the system. This means that polluted water can run right into another water source.[8]

That can mean bad news for the Lake of the Ozarks.

Because the ground has lost its shield to the crumbling rock of the karstic landscape, groundwater can easily become contaminated. And whatever comes into contact with the soil can easily find its way into the water supply. This is the concept of nonpoint-source pollution, which means it's not a direct dumping of a contaminant, but a trickle that leaches its way into the water supply from somewhere else. Lawn fertilizers and pesticides from nearby farms, plus the runoff from parking lots, all could be potential nonpoint-source pollution creators.

This is where the idea of the watershed, and how everything within and near the water body, comes into play.

The Lake of the Ozarks Watershed Alliance is focusing on landscape barriers for lake homes that back up to the water. Other groups are checking out runoff from highly trafficked areas. The idea is that, with the geology the way it is, pollution must be stopped before it meets the ground or a solution must be found that stops it in its tracks. Educating the polluters is one of the keys to solving the problem.

Scientists have followed the lake over the years and noted the roles development and the lake's geological characteristics play in keeping the area ecologically sound.

In 1985, James Vandike and colleagues John Whitfield, Donald Meier, and Cynthia Endicott evaluated the groundwater and surface-water contamination potential in a study for the Missouri Department of Natural Resources.

The study spelled out the area's geology and soil structure and how its vulnerability, coupled with the area's expanding development at the time, could prove harmful for the water sources of the region. It also discussed how prolific private wells on many lake properties functioned improperly.

It's an eerie premonition of what civic groups are pushing for now by way of organized sewer districts and calling for septic tank updates and wastewater system checks.

"The nature of the carbonate aquifer in the study area is such that any activity on the surface will have some effect on the groundwater system," the report says. "Improper waste disposal is likely to cause at least local groundwater contamination."[9]

It goes on to conclude that private waste effluent is the greatest threat to groundwater quality because of wells, septic tanks, and the area's unforgiving geologic features.

"The effluent either surfaces and flows into the lake, seeps downward through the thin soil and enters the lake downslope, or recharges groundwater through poorly constructed private wells or permeable bedrock."

Also noted is the repeated theme threading through this book: the amount of development is proportional to the amount of wastewater produced.

In 1985 when this study was released, the most intense development was within two hundred feet of the lake and researchers concluded that approximately 45 percent of the shoreline was considered "intensely developed," a term they defined as having average building spacing less than a hundred feet apart.[10]

The lake's enormous water volume would likely spread out the contamination and distribute it without a water-quality issue. But the problem is that much of the development is clustered around the coves with limited circulation, the study says.

These are the same issues causing spikes of bacteria and other quality problems cited in modern-day studies at the lake.

The study's conclusion section calls for attention to public health to ensure safety. Public water-supply districts could eliminate some pollutants, the study suggests. Development attention and improved planning might be other options.

"A more permanent and effective, but more costly approach, would be the development of centralized sewage-disposal districts," it reads.[11]

A central location would remove contamination from shallow groundwater and therefore improve the water quality. Then the groundwater would also pick up fewer contaminants as it made its way to the lake.

Still, you can't blame all of the lake's pollution problems on the geology itself.

"Most of the influence is that it is a large surface water body, and it's a function of its watershed," said James Vandike, who is now retired and an adjunct professor at Missouri University of Science and Technology. Prior to that, he spent fifteen years as the Department of Natural Resources' groundwater section chief. "Livestock is raised in the watershed and that stretches way over into Kansas. So you can't blame the karst on the surface-water side."[12]

The number of houses and their density also play a role in what happens to the water quality. Anything that can be done to decrease the amount of waste, whether directly or indirectly, can improve the water, he said.

One consideration is that from the time the lake was constructed in the 1930s through the 1960s, little growth happened.

It wasn't until the last thirty to forty years that building took off.

The houses began as vacation homes used on weekends or in the summer.

"They would have rudimentary septic systems," Vandike said.

When owners retired and some decided to move to the lake full-time, the transient population became more permanent.

That growth is continuing to present problems in the area, as historian and longtime lake resident Dwight Weaver observes.

"When our house was built in 1970, down around the foundation was chlordane and it was believed to bond with the soil and would never be a problem. Now it's migrating from house foundations into the lake coves. It wouldn't surprise me if there were also some heavy metals in some of the lake coves," Weaver says.[13]

Chlordane, that toxic liquid insecticide, is no longer approved for use in the United States.

Another leading researcher of the lake's scientific properties is Dr. John "Jack" Jones of the University of Missouri. Jones is a Dunmire distinguished professor of fisheries and wildlife and chairman of the department. His decades of data show a history of the water's limnological characteristics. The individual measurements might not mean much to a tourist looking at whether the lake is all right for recreational uses of fishing and boating.

In the long run, though, his and his students' work can shed light on the lake's health.

Jones sits in his university office with the comfort of a scientist who has achieved the post of department chair in his more than thirty years.

His research of the Lake of the Ozarks is likely the longest running and most comprehensive.

He's published scientific papers dating back to the 1980s, and each year sends students out for more collections. His samples are related to the lake's limnology, and focus on nutrient levels of nitrogen and phosphorus and how they relate to the lake's well-being. The tests note the clarity and the plankton, the physical and biological characteristics. He started his data set in 1976. Each year they collect four times at eight different sites.

Modern studies that Jones has overseen can be used for establishing limits for nutrients, such as the standards that the Missouri officials are attempting to set.

An inventory of data from a 2008 study revealed that variables from land types and hydrology affect the amount of nutrients that go into a reservoir. The study concluded that setting standard values on nutrients could be unattainable, and each region and reservoir should be considered individually because of different factors.[14]

Jones is hesitant to reveal what trends he's following or what kind of conclusions can be made at the lake.

"I've published a few papers and [findings on Lake of the Ozarks] is going to be one of my swan songs," he says. "I'm going to throw it on the desk when I leave."[15]

Besides offering up analyses from scientific papers and passing along anecdotes of sample collecting over the years, Jones lets the numbers speak for themselves and said he can't speak to bacteria-related water-quality issues.

"No generalizations can be made," he says. "This is a water body with strong, longitudinal gradient," Jones says. "It's an enormous watershed and has a fantastic exchange rate. And, I don't have thirty years of bacteria data."

He points to a study that shows the nutrient changes from wet and dry years.

"The whole topography is not suited for wastewater."

To measure water quality, you must first identify what the water is used for, Jones said, whether it is for domestic, agriculture, industry, fish, wildlife, or recreational uses. The landscape, climate, hydrology all play into the water's quality.

Jones, as a good researcher will do, lays out the objective scenario. He and the students he mentors are responsible for discerning patterns and causal variables. Then they can offer suggestions for limits and recommendations for criteria to government and regulatory bodies based on their conclusions.

Much of the attention to the Lake of the Ozarks is localized, recreational water-quality problems, said Tony Thorpe.

"What people really want to know is can my grandkids swim in it?" Thorpe says. And the answer is: it depends on the cove, he says.

Thorpe and Dan Obrecht, who head the Lakes of Missouri Volunteer Program, are Jones's go-to experts about how the lake is doing. Thorpe is also involved with the watershed alliance.

"It's not crystal clear like Bull Shoals," Thorpe says. "But the water is an improvement over the past." Bull Shoals is a reservoir lake covering forty-five-thousand-acres in northern Arkansas and southern Missouri, created by the damming of the White River before it flows into the giant Mississippi.

CHAPTER 7

Loving to Death

In the fall of 2005, Lover's Leap—that spot high above the Lake of the Ozarks with the view of converging Osage and Niangua Rivers—came into the news, yet its entry and exit were subtle.

Real estate agent Jennifer Schanuel brought the property before the Camden County Planning and Zoning Commission in September 2005. The group Villanueva/Crumby Investments had acquired the land in 2001, according to the Camden County Recorder of Deeds office. The request was for the property, which was then zoned R-1—low-density residential, to change to R-3—high-density, so that condominiums might be built there.[1]

As the commission chairman at the time, Jim Dickerson, recalls, this action was at the property owner's request. The 23.5 acres is located on Pier 31 Road.

The board took on the proposal during a busy time. Building around the lake was booming, with nearly six thousand condos under construction at one time. Sometimes hundreds of people packed public meetings to comment or protest.[2]

During public comment, Rex Derringer, who owned adjacent property, spoke about all the condos already approved at the lake and said that switching the zoning of the Lover's Leap property would be a serious mistake.

Another person testified that a nearby road was already dangerous enough without added traffic that such a development might bring.

The commission postponed a vote.

At the October meeting, minutes indicate that commissioners again delayed a vote to ensure proper notice of the meeting was given and to perhaps investigate the area's historical significance and whether it should be made into a county or state park.[3]

Real estate agent Jennifer Schanuel was trying to sell the Lover's Leap property in 2012. The county planning and zoning commission had denied historical consideration and approved it for condo development seven years earlier. This is the sign as it appeared from the water. *(Photo courtesy of Traci Angel.)*

There was much discussion, Dickerson recalled, mostly about the historical relevance and usage of the surrounding property. But the county didn't have regulations that addressed historical significance, Dickerson said. He said that previously he had urged other county officials to set up regulations to consider historical aspects of properties, as these issues were dear to him as a descendant in a long line of area dwellers and former president of the Camden County Historical Society. The county commission dismissed the idea, he recalls.

However, the commission still approved to move the request along with a six-to-four vote.[4]

In the end, the commission denied the historical significance at a meeting on December 21, 2005.[5]

A day later, the commission approved the land for condos.[6]

The idea to protect Lover's Leap came up again four years later because overuse had bruised the landscape, as is often the case with iconic places that offer a breathtaking view and beckon foot traffic.

In the summer of 2009, people who had walked onto the private property caused owner Pat Crumby to barricade the area because of the trash and litter people had left behind.

Crumby and her family had allowed people to hike to the well-known spot. But she told the *Lake Sun Leader* that she often found discarded garbage.

"It's such a beautiful spot and it would really make a beautiful park area," she told the newspaper. "I wish people would stop doing it. It is ruining one of the most well-known landmarks in Camden County. I don't mind people using it but they need to take care of it and not leave their trash and junk there. It isn't a dump."[7]

The newspaper posed creating a park to readers. They received varying opinions, including a no from James Hall in Camdenton. He reasoned it was private property.

"She has every right to gate it off if the trash problems persists." The state should not have the site, he said.

On the other hand, Mark McBride of Illinois and Sunrise Beach said he would rather it become a park than condos.

"A park area would be much better than condos. I have property near Lover's Leap and have been afraid for the past several years that a developer would buy the land, denude it of all vegetation, and build condos, forever ruining the landscape," he said.[8]

CHAPTER 8

Documented Pollution Past

Arlene Kreutzer once took boat rides across the lake and marveled at the trees, the open land, and water. She and her friends packed a picnic lunch and looked to the rocky bluffs and the inviting shoreline along the way to their destination. "Special places," she calls them.

She and her husband fled Kansas City for a quieter life to live among nature after he became ill from heart trouble. They operated a resort called Ro-Anda Beach in the 1970s. She later worked for the chamber of commerce.

Kreutzer misses the "nature of it."

"It's grown from a resort area to a general city atmosphere, and a lot of the beautiful things we saw around the lake are covered with resorts and lodges and houses. It's just changed," she laments.[1]

Friends of hers have moved away, upset about the overgrowth and nostalgic for how the area used to look. Now, it's dock after dock along the waterfront. It's maddening, boats going up and down the shoreline, stopping at the bars and restaurants.

"The beauty of the area is gone with all the construction, and it's [that beauty] that I really miss," she says. "It's too commercial and too inhabited."

Warnings of what could happen started decades ago in the forms of studies and conclusions—academic, scientific, and civic. They continued throughout the years. They spotlighted what action could be taken, often in the form of organized utility and sanitary districts. Their summaries also noted what could happen if development came without planning.

Yet years later, after the studies' conclusions were drawn and warnings blared, the lake community rests in the same predicament:

community members are still trying to organize sewer districts, and inspectors continue to locate culprits who dump untreated wastewater into the coves.

Ward Gifford, who served as the region's first chamber president, had called for sanitary plans and other restrictions before other members of the organizations booted him out in the early 1930s shortly after the lake was formed.[2]

The next forty years proved a time of calm where the lake drew visitors, but only seasonally, and only for a short time. Vacation homes popped up, but the local population remained small until the 1980s and 1990s. The US Census Bureau reported that Benton, Camden, Miller, and Morgan had a population of 64,546 in 1980; 77,628 in 1990; 97,101 in 2000; and 107,189 in 2010. The population was expected to continue to climb.[3, 4]

Jeffrey Mitzelfelt's thesis, which created a foundation for the modern-day bacteria sampling of coves, was published in the mid-1980s (see chapter 5). His conclusions in the 1985 thesis discussed how the clustering of buildings and the long lines of automobiles during peak times had some correlation on the sewage output and bacteria numbers.[5]

That same year James Vandike and colleagues published their report on groundwater and surface-water contamination potential because of geology and the risk of failing septic tanks, which is what many homeowners have.[6] (See chapter 6.)

The efforts were mostly academic, and occasionally at the federal level with the EPA and the Army Corps of Engineers, until the state Department of Natural Resources published a report that dealt with water quality in 1992. It showed that generally the lake was healthy but, like today, that certain coves might reach the tipping point of more bacteria than standards allow. It noted the density and potential for fecal coliform bacteria in coves with higher populations and possible contamination from septic tank discharges.[7]

Water quality reports from 1992, 1994, and 1996 all mention heavy development, both residential and commercial, around the lake as potential causes for water-quality concerns.

In the 1994 report, one of the main problems expressed by the state was sewage discharge.

While much of the highly developed areas along Highway 54 have been sewered and their wastewaters are treated and discharged to the Osage River downstream of the lake, sewage from thousands of lake shore homes is not pumped uphill to Highway 54. Thus, a large amount of treated sewage is discharged to the lake.[8] Many coves have excessive algal growth due to the nutrients in sewage, but to date, no other water quality problems have been documented.

The 1996 report mentioned small wastewater facilities and improperly working septic systems as part of the algal growth and sewage issues.[9]

Then lake residents decided to step in.

In the mid-1990s, a civic group took matters into its own hands.

The Lake Group for Clean Water and Economic Development formed because the state had passed regulations for wastewater and on-site sewer systems, says Greg Gagnon, who was president of the Central Bank of the Lake of the Ozarks. The group wanted to ensure the region could meet the standards, Gagnon recalled.[10]

"It was a concern and certainly going to impact construction and development," Gagnon said.

The group consisted of developers, real estate agents, and other business folks all coming together in a coordinated effort, he said. But they also wanted to include experts and scientists to help.

The group hired a Saint Louis attorney to evaluate what they would have to do to regulate and create proper wastewater disposal at the Lake of the Ozarks.

Dan Gier served as county commissioner for Miller County for ten years and served as a member of the Lake Group for Clean Water and Economic Development. He was a member of the task force organized to address the lake's water issues.[11]

He said the group was trying to address new septic tank requirements the state had set forth. Mostly what came out of it was an educational experience as to what group was supposed to enforce what aspects of the law, he said. For example, the Department of Natural Resources oversaw some of the wastewater plants, but the local health departments were required to address septic tanks.

In response to the report's suggestion for a sewer district, voters did not pass a proposal for the town of North Shore Rocky Mount to form a district.

And, although citizens wanted to address the issue, according to

Gier, the costs and questions of responsibility divided the counties. Benton and Miller are ultraconservative, and Morgan and Camden are progressive, based on voting history, and these lines split on how to execute the ideas, Gier said.

What the group did accomplish was publishing the *Report to the Lake Group for Clean Water and Economic Development: An evaluation of the technical, political and regulatory issues regarding wastewater disposal at the Lake of the Ozarks*.

The report outlined recommendations on how to resolve clean-water issues at the lake, including formation of an executive board to supervise other studies, as well as creation of financial guidelines for using cluster and other sewer technologies. The report cites historical accounts, such as the sewer development of Saint Louis, to illustrate the lake's challenges.[12]

The problem of cholera deaths and epidemics in the populated eastern Missouri city was "sewer construction not keeping up with development," the report reads.

One key problem, which community leaders are still tackling into the 2010s, is the cooperation of political subdivisions with authority. These "have not cooperated to the point that a consistent regulation of standards, or management of these standards is in place."

The report also shifted finger-pointing from the regulatory and governing bodies back on to the lake residents.

"The fate of the Lake's economy and the quality of its lifestyle is in the hands of the public and private sectors of the Lake community—not the Department of Health or the Department of Natural Resources," the report said.

However, the report lost some credibility among community members as one of the study's attorneys, Thomas Utterback, admitted and pleaded guilty to money laundering in 1998 in an unrelated case involving $3.2 million found in suitcases in Switzerland.[13]

Next came a water and wastewater conceptualization plan that engineering group HNTB provided. Information about numbers of on-site septic tanks (fifteen thousand to twenty thousand) is continually referenced in modern reports and was included in state reports during the attorney general's meeting in 2010 on how to handle water quality at the lake.[14]

Overall, the study looked at the feasibility of creating water districts.

The report's findings and conclusions echoed information many already suspected—the lake's permanent resident population was expected to swell as baby boomers retired, and the soils were estimated to be unsuitable for on-site wastewater treatment systems in places. It also identified concerns about 180 wastewater treatment systems discharging into the lake, along with the twenty thousand on-site systems, which health officials believed in the late 1990s to be noncompliant with regulations.

The report also outlined the poor quality of the water supply. In 1998, the Department of Natural Resources showed that one-third of statewide boil orders, or twenty-six of seventy-eight, were in the four-county lake area.

Yet, the report also claimed that in the 1970s the Missouri Clean Water Commission and Department of Natural Resources set water standards that were stricter than the ones the federal government set. This decision helped to keep the lake and coves in better condition than they could have become, according to the report.

The fifty-eight-page document offers general recommendations of a state legislative measure that would be passed to handle organizing wastewater, yet it does not offer any estimate on cost to the homeowner, or an overall estimate. It outlined methods of state and federal grants (loans, revenue bonds, sales tax) and presented an implementation schedule.

Its major conclusion was that the four-county area should adopt a sewer district plan, something that lake promoters are still struggling to create.

However, the business-led effort served as a blueprint in moving forward, Gagnon said.

"People are always looking for a quick fix and always looking to improve a system," he said. "But it is obvious in our lifetime that the lake will never have a central sewer system," but a conglomeration of smaller wastewater plants and on-site septic tanks, he said. "It's a work in progress. The 1990s [group] was just the beginning, and we always need to have that goal in place to maintain our focus for great progress."

The Missouri Health Department began a monitoring program in 1997 for fecal coliforms, a type of bacterium that originate in feces, with water samples on the Osage and Glaize arms. Results showed that during the summer months that some counts in developed coves were higher than the advised standard for whole body contact. But the state shut off the study in 2001, Dr. Pat Phillips told the *Kansas City Star* in an article in late 2009. The study should have been conducted for seven years to be effective, he said. Calls made to the health department regarding this study, and e-mails sent to department officials, as well as to Dr. Phillips, who is no longer with the health department, were not returned during the reporting of this book.[15]

In 2002, Ameren hired Duke Engineering to compile a historical study of the water quality of the Osage River and the lake. Ameren was preparing its application for license renewal with the Federal Energy Regulatory Commission and needed to compile studies for its stakeholders. Much of the report was pulled together from the Department of Natural Resources and academic studies already mentioned above.

In the summary recap, the report resounded the ongoing conclusion from the studies: that the density of development around the shoreline had an impact on the water.

"The water quality issue of most concern in the Lake has been the increased shoreline and basin development and the associated increases in wastewater loading," the report says. "Wastewater discharges generally contain high levels of nutrients, chlorides, BOD (biochemical oxygen demand) load, and fecal coliforms."[16]

Although nutrient studies over the years seem to prove that levels are stable in the lake overall, "studies monitoring nutrients and fecal coliform levels within coves have found higher concentrations near areas with high use or point-source discharges, but that overall, the water quality standards are being met."

Development concerns continued into the next decade.

Also in 2002, the Missouri Department of Natural Resources released a report listed under its Geological Survey and Resource Assessment Division about water use.

One topic was recreational water use and septic systems at the Lake of the Ozarks.

"Water quality at the Lake of the Ozarks is threatened by wastewater releases from lakefront septic systems and public sewer systems. The impact of polluted waters on recreation and tourism could be damaging to the recreational economies of the communities surrounding the lake area."[17]

The report noted that high density of tanks and that some of the public sewer systems could not support the growing demand, sending raw sewage into the lake.

In 1990, the report claims, three out of every four housing units had septic systems.

Even when at full-functioning capacity, septic systems may not filter all the waste. And, "in backwater areas (such as coves), there may be limited mixing with the main body of water passing through the channel. Consequently, concentrations of nutrients may be especially high, intensifying the problems associated with nutrient build-up."

The Department of Natural Resources' 2004 Missouri Water Quality Report expressed a similar statement in its "state concerns" section, which were the same a decade previous. The report even copied the exact same language as had been used in the department's 1994 water quality report, including the exact same paragraph that was listed earlier in this chapter when discussing that year's report.

In recent years even regulated and modern sewer facilities have had slip-ups, or in the case of the city of Lake Ozark, continual problems.

In 2008, a US attorney for the Western District Court of Missouri slammed the city of Lake Ozark with a $50,000 fine for illegally discharging sewage in the lake. The city had a history of allowing overflows of sewage into the lake from lift stations. The case was settled with a plea agreement before trial, but evidence of discharges would have been presented through citizen complaints, daily wastewater logs, and police reports.[18]

The Department of Natural Resources recorded an incident in September 2007 that sent nearly fifteen thousand gallons into the water creating "a dark plume." The elevated levels of fecal coliform were higher than all levels set for swimming and recreation.

Lift station logs showed that the pumps were down and not working properly during August and September of that year.

In a separate but similar case, officials also sentenced former Lake Ozark city official Richard Sturgeon to three years' probation and a $5,000 fine for failing to report discharges.[19, 20]

"The city of Lake Ozark repeatedly discharged raw sewage into the Lake of the Ozarks, one of the largest and most popular recreational lakes in the Midwest," Matt J. Whitworth, acting US attorney for the Western District of Missouri, said in a news release. "This criminal behavior threatened the health and safety of the public. Sturgeon not only knew that the city was polluting the Lake, but failed in his duty to report the discharges to the state."

Reports over the years, through sampling studies from government agencies and public universities, show the lake is in good health in the main channel. It's the coves that are the problems. And this is where people swim and spend the most time in the water.

Grand Glaize Beach: Is It the Goose Poop?

Sign posted at Grand Glaize Beach, June 2012. *(Photo courtesy of Traci Angel.)*

It's ninety degrees on a June Monday with a warm wind sending faint relief to two dozen people who lay next to, wander out into, and bob about the waters surrounding Grand Glaize Beach, also known as Public Beach No. 2, but no longer referred to as that because of the "Number Two" feces connotation.

Weekend traffic has faded. It's the kind of summer day captured in a novel, complete with a lazy nap in a tree's shade while the sun bounces off lapping waves. Minutes pass slower here. One slides into the next one, without sense of boundary, much like how the tiny lake-area villages melt together, stitched by the oak trees consuming spaces where the water ain't.

Lake time.

A marina is off in the distance, and motorized roar fills the air. Speed boats whiz across the main channel. Kids with plastic pails fill water and dump it onto the sand, molding it into a wet mound. And the families in their minivans keep coming, lugging blankets and towels under their arms, along with straw hats and coolers.

A prominent placard reads, "Do Not Feed the Geese."

No geese in sight today, but their feces are everywhere: on the sidewalks, in the grass. It's tough to keep from stepping in it.

Another sign is more alarming. It tells visitors the water is routinely tested for harmful bacteria and at this time it is acceptable for the public. Other times, a sign says the beach is closed.

High *E. coli* counts a month earlier, in May of 2012, prompted Bob Broz, director for the University of Missouri extension agricultural engineering and water quality program, to partner with the US Geological Survey and Lincoln University in Jefferson City to run samples for bacteria and attempt to determine the source of what could be causing the high counts.[21]

The Department of Natural Resources did a preliminary testing of two public beaches before the busy season and both tests showed high levels of *E. coli*.

"The parks have been testing all their public restroom facilities to make sure nothing is leaking from there and going into the lake and there hadn't been any rain for several days, so it

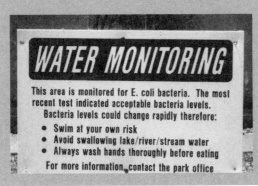

The Missouri Department of Natural Resources tested the waters weekly at the lake's public beaches during the vacation season. *(Photo courtesy of Traci Angel.)*

John Schumacher of the US Geological Survey samples the waters off Grand Glaize Beach in June 2012. *(Photo courtesy of Traci Angel.)*

seemed probable the main source of the high levels were contributed to the geese," Broz wrote in an e-mail May 17. "We went down and took samples as soon as we had heard that two beaches showed high and hopefully we will be able to see a clear indicator of what the source of *E. coli* was. The Department of Natural Resources did put up 'no swimming signs.'"

The bacteria amounts can "change from minute to minute," he said. "So what we found then might not be applicable to what we find now."

His crew took samples three feet apart and forty-five seconds apart and had significant counts of *E. coli* numbering more than two hundred colonies. This is evidence of the way water moves and the science of hydrology.[22]

Broz is waiting to hear what the source causing spikes in bacteria could be.

Earlier in the spring Broz and a Lincoln University student collected DNA samples from cattle, hogs, sheep, horses, and chickens in the lake watershed.[23]

He had helped the Lake of the Ozarks Watershed Alliance form a management plan in 2010. Now he's doing research that could shed light on the water's pollution causes and bring real answers to stop the finger-pointing. The goal is to create a library of DNA samples of different kinds of animals and poten-

Grand Glaize Beach photo, summer of 2012. *(Photo courtesy of Traci Angel.)*

tial pollutants, such as local wastewater, so that if any high levels of bacteria are detected, the DNA of the sample can be matched with the identifying source.

"It is one of those necessary things to do until you find the major sources," Broz said.

The biggest concerns in the lake is the nutrient loads, or the higher levels of phosphorus and nitrogen that enter the water sources going into the lake, and the higher number of bacteria on public beaches.

"There is so much of the local economy invested in tourism and that is why they need to keep on top of everything," Broz said.

The source-tracking project began with Dr. Guolu Zheng of Lincoln University, who requested Broz join a project that needed support from local businesses and officials. The US Department of Agriculture provided a grant to look at *E. coli* source tracking. The situation at the Lake of the Ozarks made them a viable partner for the grant. Ameren and the watershed alliance were other first partners.

The *E. coli* source tracking reads the *E. coli* found and produces something that is almost like a bar code. The specific genetic markers are matched to reveal which animal might have produced the contaminating material.

"One thing we hope to do is tell a difference between on-site sewage [pollution] and municipal wastewater," he said. "That's what we're hoping."

Broz and his group want to be able to present their data so that appropriate groups, town officials, and private citizens can put a plan into action. It may also provide information to the state health department and the Department of Natural Resources on permitting wastewater facilities or codes for on-site sewage tanks.

Once the sources are known, everyone becomes a key player, and those who have the information can take personal responsibility for things they can control, Broz said.

Volunteer efforts in controlling pollution are much better received than ones that are tied to regulations, he added.

A white full-sized van complete with filtering equipment rolls into the parking lot at Grand Glaize beach.

John Schumacher of the US Geological Survey (USGS) is here for sampling. Schumacher is the chief of the hydrologic investigations section for the USGS Missouri Water Science Center in Rolla, Missouri. Besides bacterial studies, his background includes geochemistry of springs and using borehole geophysical tools to determine sources of contamination in public-supply wells.

He is the lead investigator on a cooperative project between the Missouri Department of Natural Resources, USGS, and Missouri University of Science and Technology that is studying E. coli levels at Lake of the Ozarks State Park.

According to Schumacher, the USGS is not a regulatory agency and provides an objective and scientific voice to the data.

"The last thing I want to do is talk regulations," he says. "I'm not in public health. I'm supposed to be impartial, and I try to be independent and let the data speak for itself."[24]

He balks at calling what he does "source tracking"—identifying whether it's the geese and their poop or something else.

"It's better that if you are to prove anything, you prove what it isn't," he says. "It's awfully hard to prove exactly that it's the geese. It's easier to prove that it's not the geese."

It's nothing major to collect enough samples to get a pretty good idea of possible sources by looking at the spatial and temporal distribution of bacteria and hydrologic conditions and human activity at the time. Then later, one can do more involved testing such as source tracking using a variety of genetic techniques to see if they are consistent with suspicions, Schumacher said.

"We will determine the bacteria concentrations in the water samples first over a wide range of conditions, and then later we will look at their DNA to see if the bacteria we see are genetically consistent with what we saw in the field," he says.

On this hot summer day, the data collecting is in full force.

He lays on the dock, dips a plastic bottle for a sample. Next he lowers a water quality monitor. He records the temperature, marks dissolved oxygen, pH, and conductivity. He wants to see if all these numbers provide any kind of correlation with the bacteria counts.

Schumacher plops his bucket a few feet from sunbathers into ankle-deep water. Picnickers seems oblivious to him standing there in swim trunks, a khaki safari hat, and a white T-shirt emblazoned with Independence EcoFest, scribbling on a clipboard.

A young girl in a lime-green bikini runs and splashes nearby as Schumacher drags out water measuring instruments.

He estimates that between him and Missouri University for Science and Technology graduate student Jordan Wilson, who is working with him, they've already taken samples from the same locations thirty times this month and during the previous weekend did three sets on Saturday and early Sunday alone.

After Schumacher bends down to adjust his sampling equipment, he raises up to see someone else—another sample-taker—wading out into the place he just left.

"There's DNR (Department of Natural Resources) out to sample," he says. "They come every Monday. I didn't know what time, though."

A man in a khaki and green uniform quickly walks up the beach.

Some samples—by the end of summer as many as 1,200 samples may have been taken—will be sent to the USGS microbiology laboratory in Ohio for the complex DNA testing that is referred to as source tracking. This is a new science that is quickly evolving with techniques that are neither standardized nor absolute.

"We will be looking for genetic markers from bacteria that are much more specific [than *E. coli*] to the intestinal tracts of three groups," he said. "The markers we are looking for are geese and other waterfowl, ruminants [cattle and horses], and humans."

Inside the van, Schumacher's colleague filters a water sample, puts it on testing media, and incubates it for twenty-four hours. Another method used by the Missouri Department of Natural Resources is called Colilert and involves mixing a sample with a fluorescent enzyme that connects with *E. coli* so that after incubation the sample will glow under fluorescent light. What's causing the high number of *E. coli*? I ask Schumacher.

Hard to tell—could be geese, could be the people. It could also be raccoons, deer, or dogs. *E. coli* are found in warm-blooded animals' intestines and can find many ways to get into streams and lakes.

"One troubling thing is [*E. coli*] can live a long time outside the body," he says. They live much longer than once thought, with some studies indicating survival for months and that they might even replicate under the right conditions.

"UV [ultraviolet light] is going to kill it, but [UV] doesn't reach much below the water or sediment surface, and *E. coli* just beneath the surface can survive. It's not as easy as grabbing a couple of samples and doing DNA testing and then claiming one has identified 'the source.' Possible sources are many and can change over short periods of time. Our approach is let's take lots and lots of samples over various conditions and look around and study the parameters we collect and see if we notice patterns combined with the source tracking."

The more samples, the better.

"I think we need a lot of samples to understand what is going on," Schumacher says. "For example, there's geese excrement [in the parking lot], but if people walk or drive through it launching boats and carrying it into the water, then indirectly we are contributing with the geese."

CHAPTER 9

Modern Effort for Water Quality

Donna Swall flutters about among Camdenton area chamber of commerce members after a breakfast meeting, shaking hands and connecting people through introductions. She's brought a tableful of volunteers from the Lake of the Ozarks Watershed Alliance, the group behind the modern-day push for the lake's preservation. Among them is Mary Jo Doores, a new member who is taking notes so she can learn from Swall.[1]

"Oh, I want to introduce you to someone," Swall says. "He's helping us do something really exciting, and it's going to make national news."

Swall, who was the group's executive director, points me to a friendly looking guy. He's an engineer with Schultz Survey and Engineering, Inc.

A few minutes later Swall explains the exciting news for the spring ahead.

Rocky Mount community on the north shore of the lake is to start work for a sewer system that will connect 220 homes while getting them off septic tanks. It's the first phase of remedying the subdivision that is riddled with aging septic systems.

"We need to hook up the sewers and put in [landscaping] and filter the storm water," Swall says.

The watershed group is helping to set up low-impact landscaping during the construction process. It's a new approach that the group presented to the engineering and sewer companies that could make a big impact in the future when it comes to stopping erosion and pollution into the lake since the neighborhood has a densely populated shoreline.

Engineers are stepping forward, and the sewer districts are helping to foot the bill along with a state revolving fund (low-interest loan) to Rocky Mount Sewer District.

The project is one of the watershed group's success stories.

Swall is even more excited that previous water monitoring results can be compared with results following the project to see what kind of effect their efforts have made.

"This will tell us something," she says.

Lake of the Ozarks Watershed Alliance (LOWA) started with an attempt to bring together nearly twenty state, local, and federal entities while focusing on engaging citizens. The alliance incorporated in August 2006, adding technical advisors from various government agencies and signing on more and more lake residents to lead the way.

"LOWA is a proactive group of local residents formed to protect and preserve our lakes and watershed. Prevention is better than repairing problems economically, health-wise, and for safety," the group's website says.

The watershed group's driving force is Swall.

She has owned a home in Sunrise Beach since 1981. She moved to the lake full-time in 2006. She scuba dives and is passionate about water.

Her background in airline software marketing, organizing events, and helping to set up a watershed alliance in Kansas City that sent middle school children to monitor streams in a program called "True Blue," makes her perfect for the job.

Swall says the thousands of expected baby boomers moving to the lake could make it the third largest populated area in the state.

"We must have a plan. It's much easier to take a lake that is not in so bad shape and keep it healthy," she said just two months after the alliance's first meeting in 2006.[2]

The watershed alliance describes its mission as to protect, preserve, and improve the lake, its watershed, and natural resources while maintaining the economic, social, and environmental health of the area. The group is the main volunteer contingency for water sampling and the instigator of an effort to keep the lake clean. Its members provide documents and results for the public, nudge volunteers into coves to collect water samples, and educate the public about land management to thwart erosion.

Since its inception, it has sponsored events to create awareness for updating sewer systems and helping landowners combat pollution runoff into the lake. When the EPA put the Niangua on its impaired list, the alliance decided to craft a plan to try to tackle the problems affecting the water quality and get the lake section off the agency's targeted list.

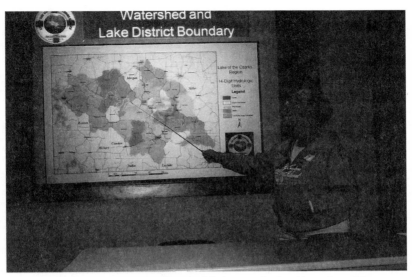

Donna Swall explains issues regarding the lake's watershed and boundary. *(Photo courtesy of Donna Swall and Lake of the Ozarks Watershed Alliance.)*

The group faces hurdles by way of the Ozarks' geography, lack of regulating jurisdiction, exploding development, and second-home population.

Yet while the watershed group is busy trying to improve water quality, they are also trying to play nice with the corporation responsible for the lake's hydroelectric plant. They partner with Ameren, which owns the lake and runs the dam. The group's grants are fueled from funding by the state's Department of Natural Resources, which was under fire in recent years for failing to immediately release lake bacteria information before a busy, tourist holiday.

The group gets along with the area businesses to keep everything in a civil and collaborative light. No one wants to say anything bad about the lake. Instead, the group focuses on prevention. They sponsor community events to educate landowners about vegetation to fight erosion. They push for water safety.

The group's intentions are true to keeping the lake clean. They have walked a narrow line of ensuring that all news they release is in good fashion with local businesses and tourism, even though they played more of a whistleblower role early on. The meeting minutes for one of the organization's first meetings recorded a conflicts-of-interest talk from attorney Melissa Manda. She explained the importance that all the board members be free from conflicts concerning what profit their involvement might bring them.

"And even more so for an organization like LOWA, just getting started, it is vitally important that not only should there be no conflicts of interest within the executive committee or Board of Directors, but in the public's eye, there should be no perception of conflict of interest,"[3] the minutes read.

In 2013, the alliance board included Warren Witt, plant manager of Ameren, the utility company.

However, the alliance is still the only organized faction pushing for change in sewer districts. Its members have even devoted time and resources to remedy signs of trouble through management plans. Their grants test water quality and then the group posts findings online for the public to see.

But are their efforts enough to keep the lake from future deterioration? Or will their voices, quieted from commercial and political interests, make them just another group of people documented among others in the lake's past who have started initiatives and watched as they fallen to the wayside?

While the watershed group is busy bettering the area, its members are quick to stay on topic regarding the water's current cleanliness and the educational efforts they are making.

One of Swall's ideas is that of low-impact landscaping—which the group has touted and held workshops on for the last several years. The idea is to plant a vegetative buffer near the shoreline to catch storm water runoff and other pollution.

"It keeps the sediment and soil on the bank," she says. There's a link to sediment carrying E. coli that could then run off into the water, Swall explains. That's why after a storm higher readings of E. coli are likely to occur.[4]

While Swall handles the marketing and business, Caroline Toole leads the environmental portion of the watershed group.

She moved to the Ozarks more than thirty years ago and claims to always have had a strong leaning toward the studies of harmony in nature. She shifted her focus to the watershed group when it was founded in 2006. Three years later she started teaching with the Missouri Clean Water

AmeriCorps program and began looking at the entire issue at the Lake of the Ozarks. And was one of the key players in drafting the watershed group's first management plan.

The 187-page document outlines the lake's history, while tracking background, population, land use, and the ecological nature of the first eighteen miles along the main channel in areas known as Lick Branch and Buck Creek. The members wrote it with the idea that the average citizen could read it and glean from it every individual's responsibility to keep the lake healthy.

Toole echoes the mantra of the alliance and area business owners: The lake is in very good shape. It is very healthy, but if we're not careful it won't stay that way. The lake needs a smart growth plan because more people are coming to the lake.

She keeps on point, despite any questions that delve into what she thinks of the lake's health.

Education. It's the key to everything, and she's quick to emphasize what the watershed group is doing to make sure the information is out there. She deflects direct questions about water quality.[5]

At a meeting in March 2012, she stressed the importance of the watershed group's landscaping plan to cut down on runoff. She gave definitions of a watershed (land that drains to a body of water).

Why care about the watershed? The lake is healthy so lake citizens should want to keep it that way, she said. In 1999, the James River in southwest Missouri overnight turned into pea soup, and the tourists went home. There were too many nutrients washing into the water. The watershed group cares about property values, swimming in the lake, and the economy, Toole said.[6]

The watershed group's new public-relations twist is to keep everything as positive as possible. That wasn't always the case. The group started out with greater transparency, welcoming many, including reporters and media to spread the word.

One member told her story in 2008. The lake drew Madeline Hutton Harrell with its promise of rustic peace. She moved from Colorado because her sister-in-law had a place there. She bought an older cabin on seven acres and continued following her interests in environmental issues.

"Things down here need to be addressed—the clear-cutting of trees was a big issue, protecting the wetland habitat, and looking for endangered species," she said in an interview in 2008. There's also little recycling, she noted.

But when clumps of algae marred her daily swims she decided to join other alliance volunteers to collect water samples.[7, 8]

"We saw a lot of water that was not clear, and I think that people that are out just racing around on Jet Skis or boating aren't paying attention to what the water looks like," she said. "The challenge is that people need to be educated about the development and the damage that is being done and how that can be reversed."

Utility company Ameren representatives attended watershed group meetings, even in the beginning.

"We went because what they do impacts us," said Warren Witt, Ameren's manager of hydro operations.

Bryan Vance of Ameren's environmental team also became involved.

Witt claims he gave Swall pointers about how to stay guarded without sounding alarms.

"I'm glad it has progressed to the level it is, but it has not been smooth sailing," Witt said in the summer of 2012. "The biggest problem has been the openness to the press—too open."

The watershed alliance needed support from the business community and had to be realistic, Witt said.[9]

"To get public support, they [felt they had to] emphasize the negative, and they turned off business people," he said. "They got people in the public involved, but they turned off the business community."

It's taken five years to overcome the barriers and the hurdles, Witt said.

"I'd say, 'Are you sure you want to say this in front of the press?' And I couldn't convince them to say it in press releases," he said. "They had to be more careful about what we say and explain the problem without exaggerating."

Goals for the watershed alliance have changed over the years, Swall said. First was to see whether a group was needed. Then other concerns, such as lake safety and a designated boat captain's program, much like a designated driver, became priorities.

Then came the group's efforts to recruit volunteers to help the Department of Natural Resources with the *E. coli* testing during its five-year study. The group's main directive at the time of this book's reporting was to keep runoff from entering the lake via the low-impact landscaping program. Members were working with nearly $1 million dollars in grant money to help with that project and others.

The topic was the focus of a July 2012 meeting with only a dozen or so attending during the busy summer season.

At that meeting, Toole regaled people with a dramatic history of the low-impact landscaping program. She started with defining a watershed program and where the grant money came from and launched into information about rain gardens and vegetated buffers. In an attempt to get others to sign on to the program, the alliance had partnered with a landscaping company that offered to discount low-impact improvement work.

"Population and population growth is part of the issue. If we get a lot of people it tips the balance," Toole said. The responsibility is everyone's, she added.[10]

The watershed group is not a watchdog now, but a community presence, Swall said.

It has also turned from a volunteer group to one with salaried employees. Swall works under contract, and Toole and Doores are salaried employees.

"The goal is not to be a drain on the community, but to have a foundation established and to keep tapping into that," Swall said.

The group's next challenge is to bring together four counties of presiding commissioners to agree on a standard for wastewater. Others have tried in the past with the same goal. It will be a clean water initiative to help those hooked up to the Department of Natural Resources permitted facilities who don't realize that they have something to maintain and to establish a memorandum of agreement for a standard wastewater management.

"The lake's footprint doesn't see county lines and what happens in Benton County can flow to Osage Beach," Swall says.

Only Camden County had a planning and zoning department; other surrounding counties did not.

"Not only do you have septics, but you have four hundred to five hundred [wastewater] facilities, condos, and houses all flushing, and some might not be working right," Swall says.

The project initially was funded by a nearly $750,000 grant from governor Jay Nixon and the Department of Natural Resources, who are working together to distribute money from the EPA.[11]

The golden project is still the sewage improvements by Schultz Surveying and Engineering. The company's goal was to start with more than two hundred homes in the Rocky Mount Sewer District and eventually service nearly two thousand customers, said engineer-in-training, Danny Roeger, who is working with engineer Jared Wheaton.

As company officials trekked through areas to sign up property owners, the real work awaits in getting the sewage to where it is supposed to go.

Infrastructure will go in and out of rock.

"It is a [time-consuming] process digging a three-foot deep trench through rock, and the steep topography adds to the difficulty,"[12] Roeger said. "The biggest problem is how tightly [close together] some of these homes are constructed. Heavy construction equipment might not fit on the property. We know that septic tanks are not the ideal resource for wastewater treatment with the soils we have here. We are seeing a number of failing septic tanks, and that is contributing to pollutants in the lake. The whole goal is to alleviate the problem and get those [septic tanks] out of there," Roeger says.

His boss, Stan Schultz of Schultz Surveying and Engineering, admits the company has a vested interest in the sewer hookup and streamlining of lake waste.

Over the years, the company's engineers have worked on projects with the Gravois Arm Sewer District, the Rocky Mount Sewer District, and the Camelot Sewer District in Camdenton. Schultz estimated they had about 1,700 customers on sewer systems in mid-2014 because of projects the firm completed. Most funding came through a variety of federal and state money.

He credits the 1999 HTNB engineering study, which the Lake Group Task Force set up, as setting the road map for future projects.

Prior to that were efforts that may have been less in the public's interest and more in a self-entrepreneur interest, Schultz said.[13]

Lake of the Ozarks Watershed Alliance Executive Director Donna Swall meets with Missouri governor Jay Nixon. The watershed group has received much funding through state-distributed grants to execute its programs. *(Photo courtesy of Donna Swall and the Lake of the Ozarks Watershed Alliance.)*

"Some of what you've had—and I'm in business to make money—is people trying to see a big windfall and spin it and sell it," he said.

The task force set up in the mid-1990s made a strong effort, but the study the effort produced left some feeling uneasy. One of the attorneys who authored the study, Thomas Utterback, pleaded guilty to money laundering in 1998.

"It tarnished the effort for a while and people were reluctant to join in," Schultz said.

He, like others, claims that the disjointed community fabric, with numerous municipalities, can hinder progress.

"It's so big and so vast that there's not one governing body, and a coalition just can't get traction," he says.

He points to Table Rock Lake and the Branson area, where he notes that Branson is a "community." The Lake of the Ozarks has grown up in a couple of generations with no organization.

"It's so big and nothing has materialized in the central management structure," he said.

The watershed group has good intentions and has been able to tap into government money that traditional sewer companies have had difficulty getting into, Schultz said. However, when it comes to testing he wonders whether they have enough samples to defend themselves statistically when speaking about the good water quality.

Business groups should get behind the watershed group's effort, which is to preserve the lake, he said. He added that the endorsement would send the right signals for promotion and tourism.

In a letter to the Lake West Chamber of Commerce, Schultz encouraged businesses to back the watershed group.

"I feel strongly that LOWA is the organization that can organize and promote a lake-wide sewer authority that will be able to promote special legislation to make the Lake eligible for large government appropriations to start building a system of 'hub and spoke,'" he wrote. "The positive publicity a united effort will create will help tourism at the Lake and promote good will with our regulators."

Most businesses are owned by entrepreneurs and they tend to not be environmental wackos or tree huggers and tend to be hard-nosed, Schultz said.

"LOWA is not a tree-hugging organization," he said. "It is progressive and it needs to do a better sales job that they aren't tree huggers and get that message to the business community."

Some in the business community didn't appreciate the watershed group's initial efforts as public watchdog. They didn't want the bad numbers and the bad press. So they developed a message of their own and formed their own group to preserve the lake, or at least its good name for the sake of business.

Many are part of the Citizens for the Preservation of the Lake of the Ozarks group with their own scientific studies and agenda. The group's main concern seems to be in the short-term, in that the tourists keep coming and spending money. The watershed alliance is a longer-term effort with watershed practices and plans and people applying for federal grants.

Although the citizens' group and the alliance are two separate groups they have some of the same players—business owners and Ameren.

"Our focus is that everyone understand the facts," the late Greg Gagnon, a businessman and leader, said of the group.[14]

Jim Divincen helps lead a group called Citizens for the Preservation of the Lake of the Ozarks, which also follows water quality issues. The group was formed with keeping lake businesses' priorities in mind. *(Photo courtesy of Traci Angel.)*

What he and other business-minded people wish to do with the group is get accurate, scientific information out that the lake is of good water quality, he had said. One of the group's efforts in 2012 was focused on high bacteria readings at Public Beach No. 2. Officials had determined that the geese population was causing the high *E. coli* bacteria numbers. State officials suggested harvesting the geese, but public protest had them considering relocation instead.

Businessman Joe Roeger said the watershed group had been opportunistic to promote an agenda that something should be done about the lake. That wasn't necessary, he said.

The difference is in the groups' motivation, Ameren manager Witt said. The alliance's motivation is the individual enjoyment of the water. The citizens' group's motivation is to protect businesses and look at the short-term. They can't afford for business to suffer right now, Witt said.

To some who care about the environment, both groups are in the ballgame for something other than clean water.

And conflicts of interests are apparent. Even Witt acknowledges that the watershed group needs the utility company and its donations to survive.

Swall might argue with him, but Witt counted Ameren as "one of the largest money providers to LOWA." He's even made copies for meetings of alliance information, he said.

Then there is Schultz, whose engineering company is gaining business after the watershed alliance secured government funding to rebuild sewer systems.

Others are watching.

"Part of the problem is the watershed alliance is composed of people who have commercial interests, and they don't want to do anything that will harm certain interests," says Ken Midkiff of the Sierra Club. "LOWA is playing an insider game."[15]

Where Are the Environmentalists?

From her balcony, Barbara Fredholm can look across to Ha Ha Tonka State Park. Her home is spacious and modestly decorated. It's inviting.

Fredholm grew up in nearby Lebanon, a mere twenty-three miles from the lake. She'd often come over on the weekends with friends who had cabins. She spent Senior Skip Day in high school down by the dam.

"It always seemed like a magical place," she said.[16]

Fredholm married at nineteen and lived in California and raised alfalfa hay. She later moved to a cattle farm in Richland, Missouri.

She and her husband decided to sell the farm and moved to the lake area, to Camelot, in 1974 for the good schools.

Her environmental leanings come from her time as a farmer and learning to love the land. A background in social work also helped her become more socially aware, she said.

The couple delved into real estate at the lake. Fredholm hesitates to give full opinions for fear she might offend a business associate. Her husband, Bob, had died a decade before I met her.

Today she checks in on the shopping center her husband built. They did try to consider the impact when building, she said. They tried to keep trees, prevent runoff, forgo fertilizer, and leave yards natural.

A few years before Fredholm contacted the Sierra Club to see if they might want to start a chapter on the lake. She heard what I was told: "The Lake of the Ozarks is a lost cause." That brush-off irritated her more.

She called a meeting and invited others. She had picked up coffee mugs and cloth napkins for the event so the group would leave without trash.

Although she claims the meeting went well, nothing ever came from the group. Fredholm said she was hoping someone would come forward to take the lead.

She joined the Lake of the Ozarks Watershed Alliance and took to kayaking and headed the watershed group's Paddle Club.

Her goal for the club is to show others the lake by kayak and experience the intimacy from a float rather than on a large boat.

Trash, wood, bluffs that were blasted away. Fredholm wants them to see it all.

Fredholm is an exception in the lake area. Few would call themselves environmentalists looking out for the lake. Even regional environmental groups steer clear of the lake.

"Environmentalists who know water quality don't go to the lake because it is such a mess," said Kat Logan Smith, Missouri Coalition for the Environment executive director. "There are so many powerful commercial interests, and they can't make it clean."

Smith's main focus is cleaning up water throughout the state, including putting in place a standard for nutrient content to prevent impaired waters.

"When you don't set standards, and apply standards, there's no place to go but downhill," Smith said.[17]

While Smith looks at systemic issues, she leaves the heavy lifting to the watershed group.

"We know the problems. They have karst geology, and it's not suitable for septic," she said. "There's a history of avoiding regulation and doing whatever, and, there's poo in the lake. A lot of us think of it as a lost cause, and if it doesn't matter to the people with a financial stake, then why should it matter to me."

CHAPTER 10

Challenges in Regulating Pollution Sources

Wesley Nyhaug is one of the responsible septic tank owners. He has made the lake his permanent home since 1998, and about ten years after he moved there he spent $10,000 to replace his septic tank.

It wasn't mandatory. The other system had been there since 1972, and Nyhaug knew it was time for a new one.

"I just thought it was the responsible thing to do, and everybody has to do his or her part to keep the lake clean," he says. "I don't know if it was leaking or not."[1]

The Department of Natural Resources isn't giving people around here much help, he said.

"They don't seem to know what the problem is, and it's going to take a while" until they figure it out, Nyhaug said.

In years past, it took the agency longer to report *E. coli* levels about the public beaches, but the agency has started posting findings within a couple of days now.

Something is going on, but Nyhaug can't put his finger on it.

Media reports in recent years have harmed the lake and make it seem like a "cesspool," when in reality it's just a few spots, he said. He's concerned that people think the whole lake is contaminated. There have always been rains to wash things into the lake, he said.

One of the more idealistic solutions put forth by community leaders for cleaning up the lake, or keeping it clean, is to get all four counties on board for one sewer district. This would streamline systems and ensure that all homes had waste disposal on the same grid, undergoing the same regulations, and the same checks and monitoring.

But all that has come out of the idea over the years are a few committees formed to discuss such a plan, and a lot of discourse about why it won't work.

The EPA released a study in 1974 that outlined a watershed management plan for the Lake of the Ozarks. It touted the plan as "the most advantageous and economically feasible plan for protection and enhancement of the Lake of the Ozarks' water quality." The study's summary highlighted an ongoing conversation: the absence of an overall system to collect and carry wastes to treatment sites.

"For implementation of an effective area-wide wastewater management system, various local governmental administrative problems must be solved," the report said.[2]

Even in the 1970s—decades before this book came out—the EPA was trying to find the solution that continues to haunt the area, as shown in the report.

> Most of the Lake shore residents occupy homes on individual lots in unincorporated segments of the Lake of the Ozarks Study Area. With the exception of very small, incorporated areas, the agency responsible for governmental functions in the Region is the county. Since only a small portion of each county is involved, and many lot owners are nonresidents who work away from the area, the county governments have not been compelled to furnish the normally expected utilitarian services. Those include water supply, streets and roads, sewerage, lighting, zoning protection and, of course, police and fire protection.

The EPA study proposed a system, new at the time, which was completely pressurized rather than the conventional way, which is pumping using the force of gravity. The operation was proposed for use near residences and businesses. It wasn't that long ago that it was common and acceptable for homes' sewage systems to just drain directly into the lake.

And, in some places, it's still happening.

The dumping is no longer legal, but reports about leaks or failing wastewater systems are a suspected underlying culprit of lake water quality.

Thousands of homes still maintain a septic tank; many homes are now connected to systems. The area's two largest municipalities—Osage Beach and Lake Ozark—built treatment plants and systems in the 1980s.

But problems can still happen. Heavy rain flooded the system in November 1996 causing seventy-six thousand gallons of raw sewage to spill into the lake near the Grand Glaize Bridge. Osage Beach reported nearly eighty spills in the three years prior to that.

A state law that became effective in 1996 required a permit for installing or repairing on-site sewer systems. A few years ago, an engineering report still indicated that the majority of the fifteen thousand to twenty thousand septic tanks were not in compliance, a number that continues to be quoted when speaking about failing septic tanks, and provided a rough estimate to illustrate the problem.[3, 4]

National Pollution Discharge Elimination System (NPDES) permits, which are the EPA's vehicle for enforcing the Clean Water Act, are needed for all discharges. Ameren in a 2002 report noted approximately four hundred permitted discharges in the Niangua River basin and Osage River basin. Most are from housing developments and businesses, but some are from animal facilities. Many flowed into the lake's tributaries.

However, bigger discharges from Camdenton Wastewater Treatment Plant or Lodge of the Four Seasons in Lake Ozark entered directly into the lake or protected coves.

A survey from the Department of Natural Resources of selected wastewater plants discharging in the lake found that 53 percent were in compliance in 1997 and 71 percent in 1998.

Many municipalities around the lake have sewage treatment plants, but the lake's topography and design forces off-the-grid homes to operate a septic tank. Owners are responsible for upkeep and health inspectors are called only when someone complains.

Who tracks a sewage concern can be a puzzle itself.

The jurisdiction can default to the Missouri Department of Natural Resources, Ameren, the state Department of Health and Senior Services, or the water patrol, depending on the situation.

According to Missouri guidelines, Department of Natural Resources governs wastewater treatment in residential housing developments of seven or more lots when lots are less than five acres. It also approves treatment for multiple family housing developments with seven or more units that discharges "in the subsurface soil absorption systems with sewage flows less than or equal to three thousand gallons per day." This means motels and hotels.[5]

But after the Department of Natural Resource's approval, it's up to the state health department to grant construction permits for systems.

The Department of Natural Resources then regulates wastewater flows of more than three thousand gallons daily and is responsible for condos, restaurants, and other commercial properties that discharge less than three thousand gallons daily that "do not discharge into subsurface soil absorption systems," such as lagoons.

The county health departments are responsible for checking septic tanks and other commercial, multiple-family residences, and commercial building systems less than or equal to three thousand gallons daily into an on-site, septic tank.[6]

Whew.

After the Missouri attorney general called a symposium in 2010 to discuss water quality at the lake, the office released a dozen recommendations to improve its health. Some of those addressed on-site sewage disposal systems. One suggestion was that systems must pass inspection at the time of sale during real estate transactions when the system was located within 2,500 feet of the lake. Another would give inspectors more access when they perceived a problem rather than waiting for a complaint.[7]

A few years earlier at a Lake of the Ozarks Watershed Alliance meeting, citizens asked how homeowners with leaky sewage systems might be reprimanded. A Department of Natural Resources spokesman said that it could be the municipality's responsibility if it fell in its jurisdiction. Or, it could be the Missouri Department of Natural Resources, depending on where the leakage was discovered.

Donna Swall, watershed group executive director, hopes the alliance will unite all entities to work toward a common goal.

The watershed alliance is "of the people, for the people," said Swall. "The lake is huge and there are so many governing bodies. We need a central coordinator for all these bodies. We are working with developers to protect [what we have]. We want to be a friend to everyone," she said.[8]

Businessman Joe Roeger is a guy who has been around a while. He started at the Four Seasons Resort at the lake in 1983. He then worked as a certified public accountant in Saint Louis before becoming an executive at the resort.

He became involved with development and titles a decade later and endured the recessions and booms until this day. His office along Bagnell Dam is still filled with rolls and rolls of property maps for his company, First Title Insurance Agency.

Roeger was part of a committee in the 1990s that explored the option of an all-encompassing sewer district and took a look at lake development. At the time, Roeger said, he and others were interested in a plan so they could get in compliance after the state issued new regulations for septic tanks.

They wanted to know the best way to approach dealing with the new laws.

The group disbanded because members found out there was "not enough political will or authority for a centralized sewer," he said.[9] "It didn't go very far politically," Roeger said. "It was politically impossible to get four counties to come together."

He's concluded a centralized sewer system is unlikely.

"A central sewer system would cost millions and millions of dollars and there's not enough effluent to support it," he said.

Each county should be able to handle its own waste disposal, he said.

Perhaps the best and most reasonable solution is to install a cluster system of collecting wastewater from residences and then connect them as the lake grows.

"The problem becomes who owns the land where the infrastructure is and how to gain access when needed," he said.

There's no planning and zoning in Miller, Morgan, or Benton counties and the counties need to address their own respective wastewater issues.

But you have to protect groundwater from wastewater contamination, says James Vandike, former chief of groundwater geology section of the Water Resources Center for the Department of Natural Resources.

Whenever there is as much development as is around the lake, the septic systems can affect water quality, Vandike says. From this comes another problem of who is in charge of these septic systems, and who wants to take responsibility for better practices.

"There's certainly nothing wrong with [wastewater] districts, and trying to improve contaminants, but unfortunately living at the lake means getting rid of waste uphill," Vandike said. "Most [systems] work with gravity, and that is usually why the wastewater treatment plants are at low points in the area."[10]

If you are hooking up thousands of homes and forcing sewer mains and pumping over drainage divides, it can be done, but it's not going to be cheap, Vandike said.

Warren Witt of Ameren, a member of the Lake of the Ozarks Watershed Alliance, was charged with helping organize the four-county task force for the group.

In the summer of 2012, he said that the task force had slowed in its pursuit, partly because of the bad press fallout from the water quality.

Because of the hype, businesses were anti-everything, including coming together to try to put everyone on the same sewer line, he said in an interview.[11]

He also wanted to set the record straight. He's read the water reports. He attended the attorney general's symposium regarding the lake's water quality. There's not a one-size-fits-all combination for how to fix septic tanks and sewers. The watershed group had suggested one sewer district for the whole lake, he said.

"That's not gonna happen," he said. "It's not gonna work."

In the last three decades, the larger towns have formed their own districts [Camdenton and Osage Beach] and that is helping, he said.

"As the community grows, there will be a push to get off septics," he says.

The logistics also play a role. "It can't be all theoretical. It has to be supported by people—people who will pay the bills and taxes," Witt said.

Dr. Randall Miles, a soil scientist at the University of Missouri, started doing on-site wastewater studies at the lake nearly twenty years ago through grants and some requests.

He has offered his opinion of handling sewage in the lake area. He has done some consulting work and given advice on what to do with on-site sewer systems.

But not everyone wants to hear his answers.

Those who think everyone should be hooked up to one sewage system disagree with Miles's assessment that the best way to handle sewage just depends on the situation.

His recommendation, based on his knowledge of different sewage technologies, is that there isn't one singular way to handle sewage.

"My recommendation was that not any one technology out there is going to be a magic bullet," Miles says. "In some cases where there is dense development, then to put in a system might be quite viable and one of the top choices."[12]

In other areas, without the same density of development, that kind of technology might not be justified or feasible to hook the property up to a system. An on-site system might be more appropriate.

Overall, the water quality picture is not that bad, with the caveat that the coves and the densely developed areas might have some water quality issues, he said. Some of the older cottages and houses pipe sewage into the lake, and that's where the trouble lies, according to Miles.

"The saving grace from the big picture is that the water turns over its volume so, from a larger picture, it's not that bad," Miles said. There is some karst, and the bedrock is fractured, and it's looser, and once water or anything that water hits permeates, it becomes a conduit for the water.

The other factor is that older structures were put in as weekender residences with infrastructure that might not support people living in them continuously. Now it's not just a weekend or second home, but is being used full-time.

What was appropriate for use for twenty years, solely on the weekends, no longer works today. Washing machines and plumbing lines are working overtime, and any malfunction means leeching into the ground and perhaps the watershed, he said.

And, ultimately, the funding to make it all happen can be a factor. State agencies often lack the resources or employees to keep up with sewage systems that are out of compliance with state standards.

Even environmentalists are sympathetic to the state's lack of resources in enforcing regulations.

"[The Department of Natural Resources] doesn't have the money to hire people to do the job," says Ken Midkiff of the Sierra Club.[13]

CHAPTER 11

Lake Water Quality Gets Political

Memorial Day kicks off the summer season at the Lake of the Ozarks. Families pile into their minivans and travel down from Interstate 70 that bisects the state ending in Missouri's bigger cities. Boats hitched to sports utility vehicles weave through the hilly roads from Arkansas. Shops that barely got by on slower, local patronage await throngs of customers who come to the lake for getaways on water and sand.

Anticipation for the tourists expands as days increasingly grow warmer, and the school year comes to a close.

The 2009 summer season was expected to be just as busy as in previous years, with vacationers coming for Memorial Day weekend and the rest of the summer months.

However, those out on the water that Memorial Day were without information from a recent report that showed high levels of *E. coli*. Swimmers bobbed, jumped off boats, and splashed in waves as the Missouri Department of Natural Resources kept the report from the public until after the busy tourist weekend. Those swimmers might have become sick from *E. coli*, a bacteria that can cause flu-like symptoms if swallowed or if it comes into contact with open wounds.

Department of Natural Resources spokeswoman Susanne Medley acknowledged the agency waited to release the information a month later for fear people would panic.

"We wanted to make sure we understood the problem," Medley told the *Kansas City Star* in a July 16, 2009, article. "Business and tourism was a consideration."[1]

Volunteers from the Lake of the Ozarks Watershed Alliance had helped with the collection of samples that had produced the higher numbers, which were attributed to heavy rains washing soil and other *E. coli*–laced matter from wildlife and animals into the water.[2]

This is a historical photo of Public Beach No. 1. In recent years the Missouri Department of Natural Resources tested its waters for E. coli and would close the beach if counts were high. State lawmakers approved a measure in 2013 to eliminate beach closings for E. coli. Instead, the state agency posts a sign that swimming is "not recommended" when levels are higher. *(Photo courtesy of Missouri State Archives.)*

The cover up of failing to disclose the crucial health information for fear of losing tourism dollars went straight to the top, to Governor Nixon's office, which oversees the agency. Then, fingers pointed and heads rolled.

Missouri attorney general Chris Koster said that it was Joe Bindbeutel, deputy director of Natural Resources, who chose to withhold the information. Bindbeutel told the press, nearly two months after the incident occurred, it was his choice. Meanwhile, Republican lawmakers called for an investigation.[3]

Mark Templeton, department director, was later put on leave and then went to work with BP after the massive oil spill in the Gulf.

Natural Resources officials defended their decision by saying that the department was waiting on more rainfall data because spikes in E. coli often are associated with heavy rainfall that washes the bacteria into the water.

The blame game continued for the rest of 2009. A former state official claimed the governor's office knew about the bacteria back in May.

The governor called a huge press conference vowing to improve water quality. Nixon also revealed that Department of Natural Resources failed to close beaches earlier in the summer after high *E. coli* readings.[4, 5]

Missouri state parks environmental section head Jim Yancey lost his job because the beaches remained open after he received an e-mail about the bacteria while he was on vacation. Natural Resources also released other longtime staff during the fallout.[6]

Amid the hubbub, water quality expert, Ken Midkiff, of the Sierra Club, filed a complaint with the state attorney general's office.

Midkiff said that agency violated the state's open records law. The watershed alliance members and others had asked that Department of Natural Resources release the data. Democratic attorney general Chris Koster's office said nothing illegal occurred because Natural Resources officials did not interpret the requests as "sunshine" requests, the name for the state's open records law.

"There's no question that they sat on [the information]," Midkiff says. The area is a tremendous source of income for individual owners and the state of Missouri, but it comes with a lot of politics, he says.[7]

At the Katy Trail's twentieth anniversary celebration, Governor Nixon confronted Midkiff for his involvement with the *E. coli* scandal.

"We came nose-to-nose and he was blaming me for putting him in the position to fire Bindbeutel," Midkiff recalled.[8]

Although no one could confirm that the two were related, stories swirled that some people who swam in the water that holiday weekend came down with flu- or cold-like symptoms that could be linked to *E. coli*.

The gaffe fueled Republican legislators, who hoped to point fingers at the Democratic governor's association to the incident. The lieutenant governor Peter Kinder joined a meeting of the lake's chamber of commerce members to discuss the public-relations challenge of the having to close beaches due to bacteria.[9]

In response to the media maelstrom that ensued, Koster called a symposium to bring together lake stakeholders, state officials, and scientists to take a better look at the situation. The meeting, August 17 and 18, 2010, was to address the water quality concerns and served perhaps as a chance for the Nixon administration to redeem itself.[10]

What resulted was the attorney general's office releasing a dozen recommendations that could help with long-term quality issues, including more monitoring of septic tanks, upgrades, and creation of a four-county sewer district.[11]

Nixon's office then asked for the Department of Natural Resources to check in with more than four hundred sewage facilities to see if they were in compliance. Results included eighty-two violations and forty-eight warning letters. His administration asked for legislation that would give state inspectors greater authority to take action in water bodies that are found to be distressed. The proposal allowed them greater leeway to deny sewage permits, too, to prevent further leaking problems.[12]

Koster's office boasted two years later that the recommendations and intervention did some good. A Lake Ozark enforcement initiative targeted pollution cases, for instance, to ensure they stayed in compliance. The legislature passed a bill that required the state health department to come up with a tool for measuring bacteria in the beach areas, and the department was instructed to hold meetings to pump up the septic-system inspections.[13, 14, 15] Ameren agreed to fund five more years of water quality testing. And the state awarded the Lake of the Ozarks Watershed Alliance nearly three-quarters of a million dollars for its erosion-fighting landscape program.

Earl Pabst was one of the Department of Natural Resources employees who became tangled up in the communications and political mess resulting from the post-Memorial Day release of those test results.

At the time, Pabst was the department's deputy division director of the environmental quality division. His division was responsible for all environmental programs, including drinking water, hazardous waste, and air pollution.

He had worked for Natural Resources for thirty-five years. He retired shortly after the legislative investigation into why the department withheld the release of the *E. coli* data.

In 2013 he was working for a consulting firm in mid-Missouri doing environmental related work that focused on helping smaller communities tackle wastewater challenges.

Nearly four years later he spoke matter-of-factly of what happened in late May–early June of 2009. He sticks by his previous statements in news articles that he advised the department's management to release the data. He said he spoke with deputy director Joe Bindbeutel and department spokeswoman Susanne Medley.

"The data results were high, but we expected them to be because of the rainfall event just prior to the sampling," Pabst said in the summer of 2013.[16] "Runoff from the rainfall resulted in higher results as they would in any lake or stream."

What got lost amid all the political finger-pointing and newsbytes was the purpose of the monitoring program that produced the higher numbers.

The program was part of a five-year study, and the data was to be calculated among the study's conclusions. The numbers were never part of the ongoing monitoring that checks regularly for water health at the state's public beaches.

"It was designed as a look at overall water quality as it relates to bacteria," Pabst said. "The fact was that [this intention] was lost almost immediately with the media and politicians. They never quite grasped that."

But that's not an excuse for not releasing the information before the holiday weekend, he said.

Pabst said he isn't bitter about what happened.

"It's not the way I intended to end my career of thirty-five years," he said. "I had planned on working there until I retired. It's just unfortunate that people who knew otherwise couldn't step up and take responsibility."

Pabst said that Natural Resources has a lot of good, dedicated people intent on protecting the environment. They are "really dedicated state employees," he said.

Chad Livengood was a reporter for the Springfield *News-Leader* during the controversy. When he learned that spokeswoman Susanne Medley said that the results were held because of the tourism holiday, he decided to take action.

"It just kind of smelled like there was something there," said Livengood, who was covering the Michigan Legislature for the *Detroit News*, in the summer of 2013.[17]

He filed an extensive open records request to Natural Resources for e-mails and received a large box. Livengood began chronicling how the e-mails showed that Nixon's office knew about the results sooner than they had claimed.

What was a political story for Livengood turned into a larger one with many arms, including how this gem of a lake had scattered consolidated sewer systems and many other problems, Livengood said.

He asked to join health inspectors and others to see the problems for himself.

"I find it absurd not to have a coordinated sewer system," Livengood said.

Pabst also pointed to what he believed as the lake's overall solution when it comes to clean up.

"In my opinion, now and back then, there is a proliferation of wastewater treatment facilities and failing on-site systems," Pabst said. "My recommendation back at the public meeting was that the counties should begin working together to develop a regional sewer system and promote cluster systems, and target failing on-site systems."

After the fallout, the Department of Natural Resources tried to polish its image. The website displays updates on all testing. And the first two links on the home page highlight "water quality" and "Lake of the Ozarks water sampling."

But "Poogate," as one blogger tagged it, had shaken the governor's office and put the Department of Natural of Resources on edge when it came time to contact them for this book.

Warren Witt, hydro-operations manager at Ameren, defended the department's delay in releasing the results. He said the results could have been misinterpreted by those unfamiliar with the situation.

"They were protecting people from their own ignorance," he said.[18] "They are victims of public opinion and are good people trying to do a good job. They are scientists dealing with facts."

Contacting scientists for information during the writing of this book often ended with referrals to the department's public-relations director. One Department of Natural Resources affiliate said that he wanted to talk to me, but that the department, since the delayed release of *E. coli* data until after Memorial Day weekend 2009, had many scared. No one wanted to say the wrong thing, and the department was carefully monitoring who spoke to whom.

When I started making requests for public records and reaching out to the public-relations office to answer much-needed questions for the book, I encountered many questions: What kind of book is it? Will it be science driven? Who is the audience? Will there be photos and graphs?

I obliged by sending some information to open the door and foster a working relationship. This didn't happen.

The following are e-mails taken from an exchange with Renee Bungart, Department of Natural Resources public-relations director. I had a few questions to ask of department scientists and was hoping to receive some deeper explanations from the studies.

Bungart had helped me in 2008 when I was investigating the issue for a magazine story, but said she could not help me this time because interviewing for my "for-profit" book was not the best use of the agency's time. I also found myself reporting in an election year in which Governor Nixon was seeking reelection.

From: traciangel@hotmail.com
To: renee.bungart@Department of Natural Resources.mo.gov
Subject: Questions for Tim Rielly
Date: Wed, 28 Mar 2012 13:27:21 -0500

Hi, Renee -

These questions are for Tim Rielly regarding Rebecca O'Hearn's master's thesis (2009) from the University of Missouri. I am attaching a copy for reference, if needed.

Please let me know if you need something else from me in order to have them answered.

Thanks for your help.

Best,
Traci

You helped with Rebecca O'Hearn's results for her thesis? What did you do? Run the samples for the data?

After her findings, O'Hearn concluded that the Lake of the Ozarks' health overall was good, but some of the coves lacking consistent flushing rates could be subjected to E. coli during rain events or to faulty septic systems. Was this accurate?

O'Hearn also identified a cove near the Grand Glaize Bridge and State Park with elevated numbers. Is Department of Natural Resources (or anyone) following up with testing for this? Is this the study that is awaiting summary at: http://www.Department of Natural Resources.mo.gov/loz/index.html

Do you know much about DNA tracking? Would this help to answer some of the questions about sourcing and causes when receiving samples with elevated bacteria numbers?

Thank you.

Unfortunately, the nonresponse became the story. Six weeks after reaching out to Department of Natural Resources through its press officer to seek permission to speak with one of the department's scientists about a study, I received this e-mail:

From: renee.bungart@Department of Natural Resources.mo.gov
To: traciangel@hotmail.com
Date: Mon, 7 May 2012 12:58:21 -0500
Subject: RE: Questions for Tim Rielly

I apologize for the delay in getting back with you regarding your request to interview staff. Upon further discussions with involved parties, we have determined that we will not be able to fulfill your interview requests. We understand that the University of Arkansas Press has contracted with you to write this book, which will be sold for a profit. Given that we are a public agency supported by tax dollars, we have determined that it is not appropriate use of limited staff time to participate in these types of interviews. Again, I apologize for this inconvenience.

However, please continue to use the department's website to complete future sunshine requests. I believe you have been in contact with our sunshine request coordinator, Veronica Caldera, and she has scanned the document you requested and provided you with a link to the document online. Keep in mind that future requests may include fees depending on the size of the document(s) and amount of time it takes to complete the request.

Thank you
Renee Bungart

I attempted to go higher up, to Governor Nixon, who had vowed transparency and action when it came to the Lake of the Ozarks.

From: traciangel@hotmail.com
To: scott.holste@mo.gov
Subject: Request for help
Date: Mon, 21 May 2012 05:09:35 -0500

Hi, Scott:

I hope you remember me. You have helped me in the past on previous news reports when I was at the Associated Press in Jefferson City and St. Louis and later when I was at St. Louis Magazine.

I am currently a freelancer in Kansas City working on a book, under contract from the University of Arkansas Press, about the Lake of the Ozarks.

I have been in contact with the Department of Natural Resources regarding public records and they have accommodated my requests. However, when asked to follow up with questions to ensure accuracy and fairness they have not allowed interviews with staff because, "Given that we are a public agency supported by tax dollars, we have determined that it is not appropriate use of limited staff time to participate in these types of interviews." This response came after more than a six weeks of follow-ups and correspondence with media director Renee Bungart. Our e-mail correspondence is below.

I have reached out since receiving Renee's reply in hopes that I could work with the department on any concerns they have. I have even suggested question / answer e-mail interviews and allowing them to review information / material used so they are more comfortable since this is an independent project rather than a media-driven one. I have not heard from Renee or anyone in the department since asking to speak more about how we could work together.

My review of news articles and press releases suggests that the governor's office has vowed to be more open about the Lake of the Ozarks and this book I am writing is an opportunity to go more in-depth about all issues regarding the Lake's environmental and developmental history. I was hoping to include state scientists' and staffs' voices among all the other outsiders and analysts in the book's writing. And the Department of Natural Resources's response to my queries as a working journalist does not seem to reflect what the governor's office has reported to do in recent years.

Any insight or help you could provide would be appreciated.

Thanks for your time.

Regards,
Traci Angel

Here's the response from Holste, Nixon's press secretary:

From: Scott.Holste@mo.gov
To: traciangel@hotmail.com
Date: Mon, 4 Jun 2012 09:09:05 -0500
Subject: RE: Request for help

Hello, Traci. If this was an issue where you were unable to obtain public records for your research, I think I could try to help. Unless I'm mistakenly reading this e-mail chain, the records you have requested from the Department of Natural Resources have been provided to you—please let me know if that is not the case. Insofar as the Department making their people available for interviews, I believe the Department of Natural Resources is in the best position to make the call on that, and I won't be compelling them otherwise.

Sincerely,
Scott Holste
Press Secretary
Missouri Gov. Jay Nixon

Current Department of Natural Resources director Sara Parker Pauley also declined an interview request that was formally mailed to her office and I then e-mailed to her assistant:

Sara Parker Pauley
Director, Missouri Department of Natural Resources
P.O. Box 176
Jefferson City, MO 65102

August 11, 2012

Dear Ms. Pauley,

I am writing to request an interview with you for a book the University of Arkansas Press has contracted me to write concerning the developmental and environmental history of the Lake of the Ozarks.

A little background about myself: I am a professional journalist who has covered a wide range of topics for newspapers and magazines in Missouri and worked at the Jefferson City and St. Louis Associated Press bureaus. I have a master's in health and environmental journalism from the University of Missouri. I have been following Lake of the Ozarks issues since first meeting with Donna Swall in 2006, when she helped to organize the watershed group there.

Over the last eight months I have interviewed many sources and devoured state and independent studies, historical documentation and research data to provide a thorough and comprehensive look at the Lake over the years. Unfortunately, during my reporting, your communications director Renee Bungart has denied my requests to follow up with department scientists to clarify information or provide added context. She wrote in a May e-mail that the department had decided speaking with me was not the best use of the agency's time.

I hope an interview with you could provide a voice from the department about the agency's role in the Lake's future and what goals your agency might have for the region moving forward.

I would like to interview you in person some time in September, if possible. I understand your time is valuable and would be happy to send a few questions in advance. Please let me know if an interview is feasible and we can make arrangements.

I look forward to hearing from you and will be following up on this letter in the coming weeks.

Best regards,
Traci Angel

The Department of Natural Resources' caution shouldn't be a surprise after the criticism and scrutiny the agency faced, I was told.

Steven Mahfood, a former Natural Resources director assigned by governor Mel Carnahan in the late 1990s, spoke candidly to me about what the department is going through. Although he did not directly work on the Lake of the Ozarks' issues, he said he has a good idea why the agency is remaining mum.

Most notably was the excuse often given—a limited number of personnel to handle the many issues the agency is expected to tackle.

"Just what I can glean—as [they have acted] in other issues—is that they are so far down in the number of staff, that this is a part of their decisions," especially if it's a difficult or controversial issue, Mahfood told me in a phone conversation in the summer of 2012.[19]

"They might find themselves unable to do things [like interviews] with what they have now. And sometimes they are worried about the criticism of going outside the normal status . . . especially with the Lake of the Ozarks," Mahfood said. "They want to put that behind them."

When he was at the helm, Mahfood said he had enough problems that he could look back after the fact and question why the agency didn't do something.

"My suspicion is that they've got so much on their plates that they don't want to go back and relook [at the E. coli data and studies] and would rather leave that to the others to judge," said Mahfood, who became an environmental consultant to the Nature Conservancy and other organizations.

Budget cuts, reduced staff, and the state's limited resources are also reasons officials gave when they answered questions about where the state is lagging in environmental protection.

The EPA had been pushing the Missouri Department of Natural Resources since 2001 to put together water-quality standards for lakes, as required by the agency's impaired report.

John Moore, one of the founding directors of the Upper White River Foundation, a watershed group in southwest Missouri, said everyone looks to state agencies to provide the framework for cleaning up and prosecuting environmental offenders, but they can't do it all, said Moore.

"They don't have enough people to go out and do it," he said.[20] "This is where our watershed organizations play a vital role in engaging people. Many of those things are in coordination and in grant support. It's a complementary role of private with federal and state organizations."

CHAPTER 12

Other Regional Watershed Groups

People in the Lake of the Ozarks need not look very far for examples on how to tackle water quality.

Watershed groups just south in the Springfield and Branson areas have organized over the last decade and can tell of success stories such as the James River basin's ongoing community involvement and the water issues and challenges they face.

While the Springfield and Branson areas also must watch septic tanks, sewage lines, and runoff, they also keep an eye on point-source pollutions—the large animal farms in nearby areas.

Southwest Missouri's growth in recent years is one reason the watershed groups were formed. And with the growth came a delicate balance of environmental interests and increasing business interests.

Todd Parnell grew up in Branson and spent a lot of time on the water. When he returned to the area in the late 1990s, the reality didn't hold up to his memories of crystal clear Table Rock Lake.

"I couldn't see into it," Parnell said from his cell phone in 2008 after meeting with Missouri's Clean Water Commission.[1] "It just wasn't in as good of shape, and there wasn't attention being paid to it," said Parnell, Drury University's president and former president of Signature Bank and Truman Bank. That sparked his work with the James River Basin Partnership and Upper White River Foundation.

John Moore, one of the founding directors in 2001 of the Upper White River Foundation, became involved with the group after he retired from his career in education, during which he served for twenty-two years as president of Drury University.

"The concern was pretty straightforward," Moore remembers. The development of the region of the Ozarks that runs into northwest

Arkansas had been booming for the last twenty, thirty, forty years, he said.[2]

Watershed Committee of the Ozarks is one of the first groups to bring awareness to the issue. The committee began operation in 1984 after a Springfield city task force suggested the creation of a "permanent body whose primary purpose would be oversight and protection of public drinking water sources," according to the group's website.

Five years later, it became a nonprofit organization that joined efforts with landowners and businesses. One of its pet projects was the Watershed Education Center at Valley Water Mill Park in Springfield. The project had a host of community partners.

The $6 million price tag covered the cost of the land and funding for a main education building. The project includes restoring a wetland area, building trails, outdoor classrooms, and a building to house the Springfield-Greene County Parks Department Outdoor Initiatives/ Maintenance Building, which will have a vegetated "green" roof.

"We are hoping to use the power of persuasion and education to get people to protect our water," said Loring Bullard, the group's chief executive officer.[3]

Bullard hopes this permeates to everyone, including those who are fueling the growth.

"You have some businesses very interested in water quality, and they like to go fishing or floating, and they understand that it affects them personally," he said.

Moore agreed.

"All of us who started the conversation have a long-standing love affair, and we brought in corporate self-interest," he said.

Holly Neill, a former executive director of the James River Basin Partnership, worked with businesses, too.

The group assisted with a LEED platinum retail shopping development, Green Circle, that includes a green roof and a system to catch all the rainwater that will be used for flushing toilets and irrigation, and the pavement is sponge-like rather than non-porous.[4]

Another effort focused on more urban issues and working with Missouri State University to sample water coming off of residential subdivisions in Ozark and Nixa.

The partnership teamed up with Dr. Bob Pavlowsky of Missouri State University, director of the Ozarks Environmental and Water

Resources Institute, to look at the quality of water runoff coming off subdivisions.

Other ideas looked to address one of the area's biggest potential water contaminants comes from the large poultry farms, also known as CAFOs or concentrated animal feeding operations.

A US Department of Agriculture survey in the late 1990s called *A Cooperative Nutrient Management Project in the Upper Shoal Creek Watershed* noted the "abundance of clear streams, springs, and lakes." But the poultry industry created serious water concerns. Data collected showed that in 1996, the value of Missouri poultry production surpassed beef. Southwest Missouri's annual poultry production was approximately 190 million broilers and 20 million turkeys, and generated 475,000 tons of litter per year.[5]

Tulsa officials sued six poultry companies (Peterson Farms, Cargill, Tyson Foods, Cobb-Vantress, Simmons Foods, and George's) and the city of Decatur, Arkansas, in 2001 for polluting creeks and streams that emptied into lakes Eucha and Spavinaw, home of the city's main drinking water sources.[6]

In June 2003, Tyson officials admitted to illegally dumping untreated water from its processing plant near Sedalia, Missouri. The plea included twenty felony violations of the federal Clean Water Act, and the company agreed to pay $7.5 million for cleanup.[7]

A year later, James River Basin Partnership, with support of the EPA, began a nutrient awareness program that gathered information about how many landowners use poultry litter and how landowners would back a program that used poultry litter as fertilizer. The study in 2004 quantified and reported statistics on the challenges—poultry farms pump $1.78 billion into the area's economy—and the amount of waste produced. The study showed approximately 100 to 120 tons of litter are produced annually by one hundred thousand chickens, providing a management challenge for many poultry producers. Their study also found that 40 percent of 227 farmers surveyed used poultry litter as fertilizer.[8]

"A lot of that [study] was to see how much poultry litter was being produced, and now we have that data," Neill says. The partnership had tried to put farmers into contact with companies that wanted to get rid of the waste, Neill says.

The James River Basin Partnership, based in Springfield, Missouri, is a watershed group that hosts many preservation events, including a river rescue to clean up along the river. *(Photo courtesy of Ian Pitts, James River Basin Partnership.)*

Watershed groups can't escape the companies that often are indirectly responsible for some of their water-quality woes. In fact, they encourage their participation.

Ameren is a regular donor to activities involving the Lake of the Ozarks Watershed Alliance. One example was the sponsorship of the alliance's designated captain campaign to reward sober boat drivers with a symbolic key chain that would provide free beverages at participating businesses.

Even John Tyson, CEO of Tyson Foods, serves on the board of directors of the Upper White River Basin Foundation.

"There's no conflict of interest," Moore said. "Companies like Cargill and Tyson have a very great self-interest in keeping the waters good and clean for the people who come here. It doesn't hurt to be green today and have enlightened corporate policies."

Parnell concurs. "[Poultry companies] don't want to be perceived as polluters and work hard to improve their impact and image," he said.

Yet, limitations exist even with plans on the books.

In 2005, Floyd Gilzow, executive director of the Upper White River Basin Foundation, and local lawmakers got together to initiate what is known as the Upper White River Basin Watershed District for the areas of Greene, Stone, Christian, Taney, Barry, Douglas, Webster, Wright, and Ozark counties. The intention was to monitor water quality in rural areas with boundaries outside sewer districts. It gives the counties the authority to regulate water quality and sewage concerns, specifically to provide maintenance for failing wastewater treatment systems.

US Geological Survey water sampling stations, located on various streams and rivers, collect massive amounts of data on water quality and flow. Other sources of water data come from a variety of partners including county health departments, state agencies, universities, and federal agencies, said Jaci Ferguson, director of the EPA's Southwest Missouri office.

One challenge is identifying sources of data and then helping watershed groups and agencies use the data, Ferguson said.[9]

Moore says that's one of the lingering issues. How do you know if water quality is good or bad? Instead of anecdotes and water samples, maybe we need the equivalent of the Consumer Price Index.

"It would be systematic, like an annual state of the watershed," Moore said. "We need to do that so we have good science and good public education to understand the issue before they get involved."

Funding is another constant obstacle, as is tailoring projects where the funding is, Neill said. The James River Basin Partnership had received nearly 70 percent of its funding through grants and that's not sustainable, Neill said. "We have to keep up with growth, and it's just exploding," she said.

James River Basin Partnership is not alone.

"Watershed groups are typical nonprofits, and they chase federal grants. They are all looking for sustainable funding," said Moore of the Upper White River Foundation.

On a positive note, watershed groups are bringing a lot of attention to people so they can get their hands dirty.

James River Basin Partnership focuses on the grassroots level, "from the bottom up," as Neill put it. Its projects involve rain gardens and working with rain barrels to reduce storm water runoff. They've set examples in gathering more than one thousand homeowners to pump out septic tanks and educate them on routine maintenance and better ways for sewage disposal.

In the ideal world, everyone would put the environment as a priority and not as a reaction to a problem, Neill said.

"People are still going to irrigate their lawns until we run out of water, and we hope that it doesn't happen," she said. "We have to make natural resources a priority to protect instead of taking them for granted. That's very common around here."

CHAPTER 13

Living Together—Activists, Homeowners, Policy Makers, and Proprietors

One idea most people at the lake agree on, no matter their background, is that lake development is here to stay, and therefore water quality is likely to always be a concern.

The greatest populated area, Camden County, does have a planning and zoning commission with a long-range plan. Voters approved the organization, which formed in 1997. The entity has since looked at long-range plans for the district, which covers a designated area within about five miles of the lake. Property outside the district is not subjected to the organization's regulations.

Other towns and counties are starting to pay attention too, said Mickey McDuffey, a former resort-owner turned civic leader, who has served on the Camden County planning and zoning commission since 2006.[1] Both she and her husband had been on the lake's chamber board, and each has served as president of the Camdenton Area Chamber of Commerce.

"I don't think it's too developed, and the leaders of this community can maintain the natural, rural beauty to protect the environment and still accommodate the infrastructure, but it takes planning."

Other municipalities within Camden County—such as Camdenton, Osage Beach, Lake Ozark, and Linn Creek—also use the county's planning and zoning guidelines. Yet, an area-wide development plan remained elusive, despite modern attempts by a group called the Lake of the Ozarks Council of Local Governments. The group had developed an economic development strategy in 2012 with broad ideas and general recommendations.[2] The study noted area statistics and made abstract suggestions on growth, but focused primarily on businesses and the economy. In recent years, the group had faced criticism and was looking for a director in late 2012.

A key lake challenge is "how to maintain the rural, open feeling the lake has," McDuffey said. "The question is how do we develop the infrastructure to accommodate the increasing residential and tourist populations and protect the natural beauty of the lake? Increases in population require increases in services and business. How do we maintain the balance between nature and necessity?"

There's no turning back either.

The lake was built fifty years ago without regulations as to how the area might develop.

"The cat's out of the bag," Warren Witt, Ameren hydro-operations manager, said.[3]

That's lake author and historian H. Dwight Weaver's take, too.

Weaver, whose photos, local knowledge, and postcard collection are in high demand, doesn't pay attention to any environmental movement at the lake because his hands are full documenting the history. He speaks generally and laments that no plan is in place to keep lake development at bay. He fears it will become like a city with a lake in the middle rather that the wooded refuge it once was.

One of his wishes is to have a historical structure for each decade to show the passing of time.

"But we don't have history because they bulldoze it down," the former public information officer for Department of Natural Resources said.[4]

"The lake has been compromised," he said. "The cat's out of the bag, and they aren't going to put it back in."

Perhaps no one else sees the daily challenges more than Tracy Rank, an environmental public health specialist supervisor at the Benton County Health Department. She's charged with checking out failing septic systems and possible sewage leaks.

Two years after the state attorney general released recommendations, her job hadn't changed much. She hadn't seen a difference.

She thinks people are becoming more conscientious about staying regulated and educating themselves. But she still sees the dumping and doubts it will ever fully go away. She's pessimistic about a Lake of the Ozarks central sewer system, too, as "that would take forever and a lot of money," she said.[5]

Tracking the culprits is difficult. It can be hard to distinguish where the sewage is coming from. One recent call Rank received was about three drainage pipes spilling into the lake.

People don't realize that even if the pollution is into a stream it could get into the lake, she said.

Rank's office still sees fifty-five-gallon metal barrels used as septic tanks. Many are from former "weekend" people who have decided to move in full-time. Whereas the washing machine was doing a load on the weekend, it's now operating full-time, as are the plumbing system and toilets. The cabins weren't made for all the wear, she said.

The growth that happened at the lake snuck up on residents, said Ameren consulting engineer Alan Sullivan, whose family has been around the lake for decades.

"Think about when the time the lake was built and the mentality," he said.[6] It was built for electrical power. The recreational side was something no one considered, and nobody had a clue what this lake would become, Sullivan said. City slickers came to the lake to duck hunt. Most people thought the flooded land was worthless so they sold it cheap.

Then, hunting and fishing became more popular. Water skiing as a sport garnered attention, and the Ozark Water Ski Thrill Show drew spectators.

"Those ski shows proved it was not only a great fishing lake," but water sports brought people to visit the lake, he said.

"Nobody had a master plan, and things have evolved as economic times have changed," Sullivan says.

In early years you took a cruise on an excursion boat or rented a speedboat. Now that high-powered boat is what everyone owns. Cabins of the 1950s and 1960s grew to the mansions of today with huge projects and massive docks.

Times change.

Each township has its own jurisdictional authority rather than one governing body. The area is so immense, it spans counties and counties, Sullivan said.

It's just after ten on a steamy Monday morning in July. A temperature of ninety-six degrees registers on the thermometer, and already the sun's wicked intensity bears down. I sit in the passenger seat of Barbara Fredholm's beige Chevrolet pickup truck.

She has bungeed two kayaks in the back, and we are headed to a nearby cove. She wants to show me how they've blasted one of the hillsides to build new condominiums.[7]

The seventy-one-year-old just started kayaking four years ago. She likes the connection to the lake that comes from sweeping the paddle. The view is much different than from out of a noisy boat.

We turn on to a windy, wooded road, which has seen better days. Abandoned boats with rust are still housed here; their owners unable to afford rent or have forgotten them.

Mountains of old dock foam, discarded refrigerators, broken-down cars, and overgrown weeds nearly halt traffic as the road snakes to the water's edge.

Fredholm's son, Mike, lives near Pier 31. Getting into and out of his place, because of the road's surrounding junk piles, can be an arduous task for her son, who has only partial use of his limbs because of an accident in the early 1990s.

We open the gate that leads to the put-in for kayaks. Fredholm lathers on sunscreen and readies her kayak as I get a feel for moving the kayak around the cove. Then we take off.

Fredholm leads the way near the rocky edge as the zooming Jet Skis and roaring boats zip by in the main channel.

One spot Fredholm particularly wishes to show me is Lover's Leap—the landmark in so many photos, the destination for those who want to see the greatness of the Osage River and Niangua from a bluff.

The area is now zoned for condo development.[8, 9] A For Sale sign is clear from the water's edge. We sit in our kayaks, which are knocked about by the waves coming from the larger boats in the distance, and gaze above. I imagine the people who spent a special moment looking out at the vista, or maybe took a snapshot, then think what if this were carved out for another habitable spot.

We keep going for the latest development, a collection of condominiums called Trinity Pointe, built only a few years earlier.

Road along Pier 31 Marina. *(Photo courtesy of Traci Angel.)*

"We're going around here," she tells me and points to a bar and restaurant called Larry's on the Lake.

Our paddling efforts provide a view of a hillside demolished and dug up for new construction. The water lapping wildly, we sit in our kayaks some more while I take pictures. I gaze at the blasted bluff and think about the sights like these countering the surrounding forests that lure one to a wooded, watery escape. Downsides come along with these enticements.

Expanding populations, swelling communities, and years of lax regulation are forcing landowners and concerned citizens to take action. Watershed alliances and business groups have become the driving forces, if ever so quietly, to ensure the quality of the area's water and reputation. These organizations face obstacles—lucrative construction, slow-moving municipalities, agencies, stubborn landowners, and large-farm corporation by-products that continue to challenge water quality. Then there's the mentality that historic and iconic land, such as the Lover's Leap ridge, requires no protection from dynamite.

Barbara Fredholm paddles her kayak beneath Lover's Leap cliff and toward a developed bluff. *(Photos courtesy of Traci Angel.)*

Trinity Pointe housing development. *(Photo courtesy of Traci Angel.)*

Lake of the Ozarks State Park naturalist Cindy Hall said that park officials are always consulting with people to help restore areas.[10]

"The dam has only been there since the 1930s, and the Ozarks were formed millions of millions of years ago. Who knows what is to come?" she said. "We tend to think of things in short time periods and tend not to think of things geologically changing. What human beings do is small when compared to geological time. But we do it so fast that the species can't compete, and then we won't be around to see what happens."

The push for preservation is now.

Sundown on the Lake of the Ozarks. *(Photo courtesy of Missouri State Archives.)*

EPILOGUE

The stars seemed to align for this book's timing. Two thousand twelve—the year I did the majority of writing and reporting—marked the fortieth anniversary of the federal Clean Water Act. The year also noted the fiftieth anniversary of Rachel Carson's breakthrough environmental book, *Silent Spring*. On a personal note, I gave birth to my daughter, who made her appearance on April 22, Earth Day.

As with all projects with such a scholarly or scientific focus, this could be a jumping off point for much more about the Lake of the Ozarks region's environmental topics. It could also provide a cautionary tale for other human-made lakes and the communities that are created because of them.

The book is without the many government players who denied my requests, though I tried tenaciously to include them. This compilation also lacks information on other forms of pollution, such as heavy metals, within the lake's boundaries. My search for any studies regarding these came up empty, and I did not have the resources to conduct my own testing.

Readers will also find that this work avoided the philosophical environmental question about the good or bad of hydroelectricity, which has been argued in many different platforms.

Where this effort cuts new territory is with the hours and hours of original, modern-day reporting from stakeholders that complement the historical studies and research, woven together in one place. I hope it can bring new material to the ongoing discussion about the lake's future. My intention was to illustrate how the lake had reached this modern point in its developmental lifespan, while acknowledging and chronicling its environmental consequences along the way.

NOTES

CHAPTER 1

1. Mike Gillespie, "Lover's Leap—Myth or Fiction?" Lake Area History Pages, accessed July 11, 2012, http://www.lakehistory.info/loversleap.html, 2000.

2. Colonel R. G. Scott, *Indian Romances* (Boston: Meador Publishing, 1933).

3. Ken McCarty, in discussion with the author, January 10, 2012.

4. Cindy Hall, in discussion with the author, January 4, 2012.

5. *Trails at Lake of the Ozarks State Park,* Missouri State Parks, accessed August 6, 2011, http://www.mostateparks.com/trails/lake-ozarks-state-park.

6. *Trails at Ha Ha Tonka,* Missouri State Parks, accessed January 2, 2012, http://mostateparks.com/trails/ha-ha-tonka-state-park.

7. McCarty, discussion.

CHAPTER 2

1. *Early River Navigation,* Miller County Museum and Historical Society, accessed December 12, 2011, http://www.millercountymuseum.org/rivernav.html.

2. H. Dwight Weaver, "Introduction," *Lake of the Ozarks: The Early Years* (Chicago: Arcading Publishing, 2000), 7.

3. "Bagnell Dam History," Ameren, accessed December 29, 2011, http://www.ameren.com/sites/aue/lakeoftheozarks/Pages/BagnellDamHistory.aspx.

4. "Indictments Name Craven and Seven Others," the *Nevada Daily Mail,* June 8, 1927.

5. "Library District Walking Tour—Land Bank Building," Kansas City Public Library, accessed December 29, 2011, http://www.kclibrary.org/district-tour.

6. Gillespie, "Lake History—Dam Facts," Lake Area History Pages, accessed multiple times through June 20, 2013, http://www.lakehistory.info/damfacts.html.

7. "Bagnell Dam," Miller County Museum and Historical Society, accessed December 30, 2011, http://www.millercountymuseum.org/bagnell.html.

8. Alan Sullivan (hydro-operations manager, Ameren), in discussion with author, August 30, 2012.

9. John William Vincent, "Differences of Opinion?" the *Reveille,* March 9, 1928.

10. Leyland and Crystal Payton, *Damming of the Osage: The Conflicted Story of the Lake of the Ozarks and Truman Reservoir* (Springfield: Lens & Pens Press. 2012), 110.

11. "The Construction of Bagnell Dam: When it Happened," Lake of the Ozarks Convention and Visitors Bureau, accessed January 4, 2008, http://www.funlake.com/history/bagnell/timeline.

12. Gillespie, "Myth of the Lake's Sunken Town Sites," accessed August 11, 2012, http://www.lakeexpo.com.

13. "History of Lake Ozark/Osage Beach," Lake of the Ozarks Convention and Visitors Bureau, accessed December 28, 2011, http://www.funlake.com/history/lake_area.html.

14. Gillespie, "The Lake on Trial," accessed December 31, 2011, Lake Area History Pages, http://www.lakehistory.info/snyder.html.

15. "Luxurious Lodges on Block," the *Kansas City Star*, August 6, 1941.

16. Egan v. Unites States, Union Electric of Missouri v. Same. (8th Cir. U.S. Court of Appeals, August 9, 1943).

17. "Lake of the Ozarks," *Time*, December 14, 1931.

18. Buford Foster, in discussion with the author, July 17, 2012.

19. Weaver, *Lake of the Ozarks*, 31, 87.

20. J. Roger Snipe and Conrad H. Hammar, "Economic Aspects of Recreational Land Use in the Lake of the Ozarks area," *College of Agriculture Bulletin*, University of Missouri, no. 448 (June 1942): 3, 52.

21. Warren Witt, Ameren, in discussion with the author, August 17, 2012.

22. Weaver, in discussion with the author, August 8, 2006.

CHAPTER 3

1. "Lake of the Ozarks Now Into Its Own," *Lake of the Ozarks News*, March 15, 1932.

2. Reporting from Camden Chamber of Commerce breakfast, Old Kinderhook, January 26, 2012.

3. Jeff Green, "Lake of the Ozarks FERC Project 459 Legislative Update," Ameren Union Electric, PowerPoint, presented January 26, 2012 at Old Kinderhook.

4. H. Dwight Weaver, *History and Geography of Lake of the Ozarks*. vol. 1 (Eldon: Osage River Trails, 2005), 125-128.

5. "Lake of the Ozarks Improvement and Protective Association Now Incorporated," *Lake of the Ozarks News*, March 15, 1932.

6. Dick Fowler, "Ward C. Gifford, Ward C. Gifford Realty Company," *Leaders in Our Town* (Kansas City: Burd & Fletcher Company, 1952), 149-152.

7. "President Ward C. Gifford Writes," *Lake of the Ozarks News*, August 26, 1932.

8. "Three State Parks Are Planned for the Lake of the Ozarks Region," *Eldon Advertiser*, August 3, 1933.

9. "Ward C. Gifford Resigns as President," *Eldon Advertiser*, February 8, 1934.

10. Weaver, discussion August 8, 2006.

11. *Lake of the Ozarks Water and Wastewater Conceptual Plan*, Lake Group Task Force, prepared by HNTB May 1999, ES-2.

12. Union Electric Company, doing business as AmerenUE, Project No. 459-128, order issuing new license, U.S. Federal Energy Regulation Commission, March 30, 2007, pp. 5, 82.

13. Associated Press, "Notification System Unclear for Lake of the Ozarks Sewage Contamination," August 16, 2008.

14. *Water Quality Survey Report: E. coli Concentrations in Lake of the Ozarks, Miller, Camden, Morgan, and Benton Counties*, Missouri Department of Natural Resources, January 2012.

15. Greg Stoner, "The 5-year *E. coli* Cove Study at the Lake of the Ozarks: 2007-2011," Lake of the Ozarks Watershed Alliance minutes, November 21, 2011.

16. Ken Midkiff, in discussion with the author, January 10, 2012.

17. *Lake of the Ozarks Water Quality Initiative Report*, Missouri Department of Natural Resources, December 2009, 3.

18. Lake of the Ozarks Watershed Alliance Four County Wastewater Task Force minutes, April 19, 2010.

19. Bill McCaffree, in discussion with the author, January 12, 2012.

20. Donna Swall, in discussion with the author, August 8, 2006.

21. Jennifer Byers, in discussion with the author, March, April 2008.

22. Rick King, Osage Beach superintendent, in discussion with the author, March 2008.

23. Paulette Mitchell, in discussion with the author, March 2008.

24. Jim Divincen, in discussion with the author, January 26, 2012.

CHAPTER 4

1. Nancy Zoellner-Hogland, "FERC Decision Opens Door for Ameren," *Lake of the Ozarks Business Journal*, vol. 8, issue 7, July 2012.

2. Public comments on FERC proposal draft and Ameren's response to comments, accessed April 12, 2012,http://www.ameren.com/sites/aue/lakeoftheozarks/Documents/FERCPublicComments.pdf. 2011.

3. Peggy Crockett, in discussion with the author, August 8, 2012.

4. Warren Witt, discussion.

5. The Public Utility Holding Act of 1935, U.S. Energy Information Administration, accessed multiple times through December 7, 2013, http://www.eia.gov/cneaf/electricity/corp_str/appendixa.html.

6. "Ameren Files New Project Boundary Plan to FERC," 1150 KMRS Radio, February 1, 2012.

7. Jeffrey Tomich, "Lake of the Ozarks Boundary Redrawn to Save 1,500 Homes," *St. Louis Post-Dispatch*, June 6, 2012.

8. "FERC approves boundary for Lake of the Ozarks," press release, Federal Energy Regulation Commission, June 5, 2012.

9. "Ameren: FERC Ruling Ends Danger to Lake of the Ozarks Properties," Missourinet.com, accessed June 6, 2012, http://www.missourinet.com/2012/06/06/ameren-ferc-ruling-ends-danger-to-lake-of-the-ozarks-properties.

10. Factsheet: November 10, 2011, Union Electric Company Docket No. P-459-310/Osage, "FERC Sets the Record Straight," Federal Energy Regulatory Commission, http://www.ferc.gov/media/news-releases/2011/2011-4/11-10-11-factsheet.asp. 2011.

CHAPTER 5

1. Glenn "Boone" Skinner, *The Big Niangua River* (Cassville, Missouri: Litho Printers, 1979).

2. Weaver, *History and Geography of Lake of the Ozarks*, vol. 1, 19.

3. U.S. E.PA. Mo. 303d Decision letter to Mr. John Madras, Missouri Department of Natural Resources, October 6, 2011.

4. Gary M. Pierzynski et al., *Soils and Environmental Quality* (Boca Raton: CRC Press, 2005), 134, 141, 185.

5. *Development of Nutrient Criteria for Water Quality*, Missouri Department of

Natural Resources, report by DNR's Water Protection Program, accessed February 23, 2012, http://www.dnr.mo.gov/env/wpp/wqstandards/wq_nutrient-criteria.html.

6. Letter to Missouri Department of Natural Resources director Sara Parker Pauley regarding Missouri Water Quality Standards, U.S. Environmental Protection Agency, August 16, 2011.

7. Greg Stoner, "Management Plan for Lake of the Ozarks: Camden, Morgan, Miller, and Benton Counties," Missouri Department of Conservation, 2000, 4.

8. Rebecca O'Hearn, "Nutrients, Chlorophyll and Bacterial Fecal Indicators in Coves and Open Water Areas of the Lake of the Ozarks, Missouri," thesis, University of Missouri, 2009.

9. Jeffrey Mitzelfelt, *Limnological Characteristics of the Main Channel and Nearshore Areas of Lake of the Ozarks, Missouri*, University of Missouri, 1985.

10. Rebecca O'Hearn, in discussion with the author, February 15, 2012.

11. *Water Quality at Lake of the Ozarks*, U.S. Geological Survey at Missouri University of Science and Technology, Missouri Department of Natural Resources, accessed February 13, 2012, http://www.dnr.mo.gov/loz/index.html.

12. *Water Quality Survey—Osage Beach-Turkey Bend Area—Lake of the Ozarks, Missouri*, Consulting Analytical Services International, Springfield, Missouri, 1981, 42.

13. *Lake of the Ozarks Water Quality Management Plan*, contributor United States, U.S. Environmental Protection Agency, Region VII, 1974, 2.

14. Thorpe, in discussion with the author, January 6, 2012.

15. Thorpe, in discussion with the author, April 4, 2012.

16. Kris Lancaster, in discussion with the author, October 16, 2012.

17. "Rare, Threatened and Endangered Species," Biotic Communities, Missouri Department of Conservation, accessed February 22, 2012, http://mdc.mo.gov/landwater-care/stream-and-watershed-management/missouri-watersheds/niangua-river/biotic-community.

18. "Table BcO4: State or federal listed endangered animal species found within the Niangua Watershed," Missouri Department of Conservation, accessed February 22, 2012, http://mdc.mo.gov/landwater-care/stream-and-watershed-management/missouri-watersheds/niangua-river/biotic-community/t-2.

19. "Zebra Mussels at the Lake of the Ozarks," the *Water Line*, newsletter of the Lakes of Missouri Volunteer Program, spring 2006, http://www.lmvp.org/Waterline/spring2006/zebraussels.2.html.

20. *Lake of the Ozarks Watershed Management Plan*, Lake of the Ozarks Watershed Alliance, March 2010, 33.

21. U.S. Census Bureau, retail sales expenses for Camden, Miller, and Morgan counties, 2007.

22. Fish kills in Lake of the Ozarks below Truman Reservoir during 1978-2002, Total Maximum Daily Load Information, Missouri Department of Natural Resources, December 2009.

23. Letter to Mr. Jerry Hogg, superintendent of Hydro Regulatory Compliance, Ameren UE Osage Plant from Missouri Chapter—American Fisheries Society, January 9, 2003.

24. "Ameren to Pay $1.3 Million to State of Missouri for 2002 Fish Kill at Bagnell Dam, Under Agreement with AG Nixon," press release, Missouri Attorney General's Office, May 18, 2005.

25. *We've Taken Steps to Protect Aquatic Life*, Ameren's Corporate Social Responsibility Report, 2011.

CHAPTER 6

1. Gary R. Finni, *Environmental Inventory and Assessment of Section 10 and 404 Permitted Structures and a Fish Reproduction Study at Lake of the Ozarks, Missouri, Part 1*, submitted to Kansas City District U.S. Army Corps of Engineers, October 29, 1982, I-6.

2. Mary J. Burgis and Pat Morris, *The Natural History of Lakes*, (New York: Cambridge University Press, 1987), 5, 36, 163, 208.

3. M. L. Moore, *NALMS Management Guide for Lakes and Reservoirs*, (Madison, Wisconsin: North American Lake Management Society, 1989).

4. J. P. Michaud, *A Citizen's Guide to Understanding and Monitoring Lakes and Streams*, Publication no. 94-149, (Washington State Department of Ecology: Publications Office, 1991).

5. B. Monson, *A Primer on Limnology,* 2nd ed., Water Resources Center, University of Minnesota, 1992.

6. *Karst Topography and Nonpoint Source Pollution,* the Karst Waters Institute, accessed August 7, 2011, http://www.karstwaters.org.

7. "Karst Springs and Caves in Missouri," Water Resources Center, Missouri Department of Natural Resources, accessed January 10, 2012, http://dnr.mo.gov/env/wrc/springsandcaves.html.

8. Denise Henderson Vaughn, "Karst in the Watershed," *West Plains Daily Quill,* August/September 1998.

9. James E. Vandike et al., *An Engineering Geologic Approach to Evaluating Ground-water and Surface-Water Contamination Potential at Lake of the Ozarks, Missouri*, Missouri Geological Survey, OFR-85-26-EG, August 1985, 7.

10. Ibid, 10.

11. Ibid, 12.

12. Vandike, in discussion with the author, March 15, 2012.

13. Weaver, discussion, August 8, 2006.

14. John R. Jones et al., "Role of Land Cover and Hydrology in Determining Nutrients in Mid-Continent Reservoirs: Implications for Nutrient Criteria and Management," *Lake and Reservoir Management,* vol. 24, 2008, 1-9.

15. John R. Jones, Tony Thorpe, and Dan Obrecht, in discussion with the author, January 27, 2012.

CHAPTER 7

1. Camden County Planning and Zoning Commission, hearing minutes, September 21, 2005.

2. Jim Dickerson, in discussion with the author, September 25, 2012.

3. Camden County Planning and Zoning Commission, hearing minutes, October 19, 2005.

4. Camden County Planning and Zoning Commission, hearing agenda action, October 19, 2005.

5. Camden County Planning and Zoning Commission, hearing agenda action, December 21, 2005.

6. Camden County Planning and Zoning Commission, decision of record, December 22, 2005.

7. Joyce Miller, "Owner Threatens to Barricade Access to Lover's Leap Scenic Site," *Lake Sun Leader,* June 13, 2009.

8. "The Question: Should the Lover's Leap Become a State Park?" *Lake Sun Leader,* accessed August 1, 2012, http://www.lakenewsonline.com/opinions/x488803723/Lake-Sun-E-Board-THE-QUESTION-Should-the-Lovers-Leap-become-a-public-park?zc_p=1, 2009.

CHAPTER 8

1. Arlene Kreutzer, in discussion with the author, January 18, 2012.

2. "President Ward C. Gifford Writes," *Lake of the Ozarks News,* August 26, 1932.

3. Camden County Lake Area Master Plan, a report by engineering design company MACTEC, St. Louis and Atlanta, March 2011.

4. U.S. Census Bureau data for Benton, Camden, Miller, and Morgan counties, 1980, 1990.

5. Mitzelfelt, *Limnological Characteristics of the Main Channel.*

6. James E. Vandike et al., "An Engineering Geologic Approach to Evaluating Groundwater and Surface-Water Contamination Potential at Lake of the Ozarks, Missouri," Missouri Geological Survey OFR-85-26-EG. August 1985.

7. *1992 Water Quality Report,* Missouri Department of Natural Resources, Water Pollution Control Program.

8. *Missouri Water Quality Report 1994,* Missouri Department of Natural Resources, 20.

9. *Missouri Water Quality Report 1996,* Missouri Department of Natural Resources, 36.

10. Gagnon, in discussion with the author, June 8, 2012.

11. Dan Gier, in discussion with the author, March 22, 2012.

12. Thomas Utterback and Edward Edgerley, report to the Lake Group for Clean Water and Economic Development, May 1996.

13. Tim Bryant, "Lawyer Here Admits Laundering Money: Charges Say He Took Money to Panama, Switzerland," *St. Louis Post-Dispatch,* May 14, 1998.

14. *Lake of the Ozarks Water and Wastewater Conceptual Plan for the Lake Group Task Force,* HTNB Architects, Engineers, Planners, May 1999, ES-2.

15. Karen Dillon and Judy L. Thomas, "Cleaning Up the Lake of the Ozarks a Vast Challenge," *Kansas City Star,* December 20, 2009.

16. *Historical Water Quality Study: Osage Project,* prepared for AmerenUE by Duke Engineering and Services Inc., revised January 2002, 124.

17. Mimi Garstang, *Topics in Water Use: Central Missouri,* Missouri Department of Natural Resources, report from the Geological Survey and Resource Assessment Division, 2002, 66-70.

18. "City Pleads Guilty to Dumping Raw Sewage into Lake of the Ozarks," press release, Office of the U.S. Attorney for the Western District of Missouri John F. Wood, August 25, 2008.

19. "Former Lake Ozark City Official Sentenced for Failing to Report Sewage Discharge," press release, Office of the U.S. Attorney for the Western District of Missouri Matt J. Whitworth, August 25, 2009.

20. "Tip to EPA Website Leads to Criminal Plea by Former Public Works

Director of Lake Ozark, Mo.," press release, Environmental Protection Agency, July 31, 2008.

21. Bob Broz, e-mails exchanged with author, May 17, 2012.

22. Broz, e-mails exchanged with author, June 4, 2012.

23. Broz, in discussion with the author, January 11, 2012.

24. John Schumacher, in discussion with the author, June 18, 2012.

CHAPTER 9

1. Donna Swall, in discussion with the author at the Camden Chamber of Commerce breakfast at Old Kinderhook, January 26, 2012.

2. Swall, in discussion, August 8, 2006.

3. Lake of the Ozarks Watershed Alliance, meeting minutes, September 18, 2006.

4. Swall, in discussion, July 16, 2012.

5. Caroline Toole, in discussion with the author, January 19, 2012.

6. Lake of the Ozarks Watershed Alliance, meeting minutes, March 19, 2012.

7. Madeline Hutton Harrell, in discussion with the author, March 7, 2008.

8. Traci Angel, "What's in the Water?" *Columbia Tribune*, November 23, 2008.

9. Witt, in discussion, August 17, 2012.

10. Lake of the Ozarks Watershed Alliance meeting, reporting, July 16, 2012.

11. Amy Wilson, "Lake of the Ozarks Watershed Alliance Makes Plans for $740k Grant," LakeNewsOnline.com, accessed December 5, 2011, http://www.lakenews-online.com/communities/x13297760/Lake-of-the-Ozarks-Watershed-Alliance-makes-plans-for-740k-grant, posted March 24, 2011.

12. Danny Roeger, in discussion with the author, February 9, 2012.

13. Stan Schultz, in discussion with the author, June 6, 2012.

14. Gagnon, in discussion, June 8, 2012.

15. Midkiff, in discussion, January 27, 2012.

16. Barbara Fredholm, in discussion with the author, July 16, 2012.

17. Kat Logan Smith, in discussion with the author, January 25, 2012.

CHAPTER 10

1. Wesley Nyhaung, in discussion with the author, January 18, 2012.

2. *Lake of the Ozarks Water Quality Management Plan*, U.S. Environmental Protection Agency, Region VII, Ix, 1-2.

3. *Historical Water Quality Study: Osage Project*, Duke Engineering and Services Inc., revised January 2002 14.

4. *Lake of the Ozarks Water and Wastewater Conceptual Plan*, HTNB Architects, Engineers, Planners, May 1999, 6-2.

5. "Who Regulates Domestic Wastewater in Missouri?" Missouri Department of Natural Resources, November 2010.

6. Missouri state law regulating wastewater: http://www.health.mo.gov/living/environment/onsite/lawsregs.php.

7. *Protecting Water Quality at the Lake of the Ozarks: An Environmental Road Map for the Future*, presented at the Attorney General's Symposium, August 17–18, 2010, 34, 40.

8. Swall, discussion, July 16, 2012.

9. Roeger, discussion, August 12, 2012.

10. Vandike, discussion, March 5, 2012.

11. Witt, discussion, August 17, 2012.

12. Dr. Randall Miles, in discussion with the author, January 4, 2012.

13. Midkiff, discussion, January 27, 2012.

CHAPTER 11

1. Karen Dillon, "Missouri Agency Withheld Report of *E. coli* Reaching Unsafe Levels in Lake of the Ozarks," *Kansas City Star*, July 16, 2009.

2. Dillon and Jason Noble, "*E. coli* Report delay is faulted," *Kansas City Star*, July 17, 2009.

3. Dillon, "Former Missouri DNR Official Takes Blame for Withholding *E. coli* Report," *Kansas City Star*, September 3, 2009.

4. Dillon, "Top Nixon Aide Knew About Lake of the Ozarks *E. coli* in May," *Kansas City Star*, September 24, 2009.

5. Dillon and Noble, "Missouri Governor Calls for Action to Clean Up Lake," *Kansas City Star*, September 24, 2009.

6. Dillon, "Missouri Official Involved in *E. coli* Controversy Loses His Job," *Kansas City Star*, October 23, 2009.

7. Midkiff, discussion, January 10, 2012.

8. Midkiff, discussion, January 27, 2012.

9. "Lt. Governor Peter Kinder Hears Camden County Water Quality Concerns," press release, Office of Lieutenant Governor Peter Kinder, October, 14, 2011.

10. *Protecting Water Quality at the Lake of the Ozarks*, Attorney General's Symposium.

11. "Attorney General Koster Releases Recommendations to Improve Water Quality at Lake of the Ozarks," press release, Office of Attorney General Chris Koster, ago.mo.gov/newsreleases/2011/AG_Koster_improve_water_quality_at_lake/, January 25, 2011.

12. *Lake of the Ozarks Water Quality Initiative* report, Missouri Department of Natural Resources December 2009, 12.

13. Missouri Attorney General Chris Koster, speech on Lake of the Ozarks Water Quality, January 2012.

14. Associated Press, "Koster Makes Recommendations: AG Seeks to Address *E. coli* Problems at Lake," the *Lake Today*, January 26, 2011.

15. "Legislative Proposal Would Provide Enhanced Authority to Stop New Pollution From Flowing When Water Quality Is Distressed," press release, Office of Missouri Governor Jay Nixon, www.governor.mo.go/newsroom/2009/Clean_Water_Proposal, December 29, 2009.

16. Pabst, discussion, June 12, 2013.

17. Chad Livengood, in discussion with the author, June 15, 2013.

18. Witt, discussion, August 17, 2012.

19. Steve Mahfood, in discussion with the author, June 22, 2012.

20. John Moore, in discussion with the author, March 24, 2008.

CHAPTER 12

1. Todd Parnell, in discussion with the author, March 12, 2008.

2. John Moore, discussion, March 24, 2008.

3. Loring Bullard, in discussion with the author, January 13, 2008.

4. Holly Neill, in discussion with the author, March 27, 2008.

5. Adam Reed et al., "A Cooperative Nutrient Management Project in Upper Shoal Creek Watershed of Southwest Missouri," U.S. Department of Agriculture, Cassville, Missouri, 1998.

6. "Plaintiffs Want Class-Action Status in Grand Lake Lawsuit Against Tyson Foods," *U.S. Water News* online, http://www.uswaternews.com/archives/arcrights/3plawanll.html, November 2003.

7. Ibid.

8. Dr. Janice Schnake Greene, *Use of Poultry Litter in Southwest Missouri*, a report by the James River Basin Partnership with Southwest Missouri State University, Fall 2004.

9. Jaci Ferguson, in discussion with the author, February 15, 2008.

CHAPTER 13

1. Mickey McDuffey, in discussion with the author, November 28, 2012.

2. *Camden County Lake Master Plan*, prepared by MACTEC, March 2011.

3. Witt, discussion, August 17, 2012.

4. Weaver, discussion, July 18, 2012.

5. Tracy Rank, in discussion with the author, March 12, 2012.

6. Sullivan, discussion, August 30, 2012.

7. Fredholm, discussion, July 16, 2012.

8. Camden County Planning and Zoning Commission, hearing agenda, December 21, 2005.

9. Camden County Planning and Zoning Commission, Decision of Record, Case No. 05-458, December 22, 2005.

10. Cindy Hall, in discussion with the author, January 4, 2012.

INDEX

TRACI ANGEL is a writer and editor who lives in Kansas City, Missouri. She is a former health reporter for the *Jackson Hole News & Guide* and covered regional topics while a reporter for the Associated Press and editor at *St. Louis Magazine.* She has been following the environmental situation of the Lake of the Ozarks for several years.